T0134582

Performance Analysis of Photovoltaic Systems
with Energy Storage Systems

Adel A. Elbaset
Saad Awad Mohamed Abdelwahab
Hamed Anwer Ibrahim
Mohammed Abdelmowgoud Elsayed Eid

Performance Analysis of Photovoltaic Systems with Energy Storage Systems

 Springer

Adel A. Elbaset
Minia University
El-Minia, Egypt

Saad Awad Mohamed Abdelwahab
Suez University
Suez, Egypt

Hamed Anwer Ibrahim
Suez University
Suez, Egypt

Mohammed Abdelmowgoud Elsayed Eid
Suez University
Suez, Egypt

ISBN 978-3-030-20898-1 ISBN 978-3-030-20896-7 (eBook)
https://doi.org/10.1007/978-3-030-20896-7

This Springer imprint is published by the registered company Springer Nature Switzerland AG
The registered company address is: Gewerbestrasse 11, 6330 Cham, Switzerland

Dedicated to
Our parents, brothers, sisters, teachers, and
friends for their love, encouragement, and
endless support.
We give all thanks and gratitude to our
beloved wife and to our daughters and sons
wishing from God to protect them.

بسم الله الرحمن الرحيم
"يَرْفَعِ اللهُ الَّذِينَ آمَنُوا مِنكُمْ وَالَّذِينَ أُوتُوا الْعِلْمَ دَرَجَاتٍ وَاللهُ بِمَا تَعْمَلُونَ خَبِيرٌ "
صدق الله العظيم
المجادلة/ 11

Acknowledgment

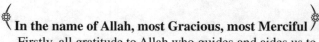

In the name of Allah, most Gracious, most Merciful

Firstly, all gratitude to Allah who guides and aides us to attain the achievements of this book.

We would like to express our greatest gratitude to the Faculty of Industrial Education, Suez University, and Minia University for helping us to finish our book. Our deepest gratitude goes to our mothers, to whom we are most indebted. We thank them for their constant love and prayers; without their prayers, we could have never been able to achieve this book. Special words to our wives for their unlimited help, patience, care, endless support, and continuous encouragement throughout our life.

Abstract

Recently, the permanent growth of the energy demand and the rapid depletion of the conventional power sources have attracted the research interests of the authors toward the renewable energy sources, especially the photovoltaic (PV) energy as alternative sources of energy. The PV energy can be utilized only during the daylight. Therefore, the integration of the PV energy and the energy storage system as the battery supercapacitor can attenuate their individual fluctuations, increase the overall output power, and generate more reliable power with higher quality to the electrical loads in the rural areas. The aim of this book is to study and design the performance analysis of the PV stand-alone systems with energy storage systems as follows:

- This book investigates dynamic modeling, simulation, and control strategy of the PV stand-alone system during variation of the environmental conditions. Moreover, the effectiveness of the implemented maximum power point tracking (MPPT) techniques and the employed control strategy will be evaluated during variations of the solar irradiance and the cell temperature. The simulation results are based on the reliability of the MPPT techniques applied in extracting the maximum power from the PV system during the rapid variation of the environmental conditions. Furthermore, it introduces a review of two MPPT techniques that are implemented in the PV systems, namely, the perturb and observe (P&O) MPPT technique and the incremental conductance (InCond) MPPT technique. The two MPPT techniques were simulated by the MATLAB/Simulink, and the results response of the PV array from voltage, current, and power are compared to the effect of solar irradiation and temperature change.
- Then, the proposed PV stand-alone system is utilized to supply the demanded power of variable loads. The PV array is connected to battery energy storage (BES) through the DC bus in order to supply the demanded power of the variable loads. Moreover, the power flow control strategy is proposed to feed the demanded power of the variable loads. The BES can act as a buffer store to eliminate the mismatch between PV power and load demand. Furthermore, the BES helps to improve the performance of the system through the control used in the

process of charge and discharge to manage the sudden load changes and helps to maintain a stable voltage level on the load and PV terminals.

- Improving the performance of the PV stand-alone system by leveraging the properties of the battery-supercapacitor hybrid energy storage system (BS-HESS), this book proposes an efficient control strategy to enhance the BS-HESS capable of the PV stand-alone system.
- The PV panels are not an ideal source for battery charging; the output is unreliable and heavily dependent on weather conditions. Therefore, an optimum charge/discharge cycle cannot be guaranteed, resulting in a low battery state of charge (SOC%). Low battery SOC leads to sulfation and stratification, both of which shorten battery life. A control strategy is essential for the BS-HESS to optimize the energy utilization and energy sustainability to a maximum extent as it is the algorithm which manages the power flow of the battery supercapacitor.
- Performance analysis of the PV stand-alone system with BS-HESS during the high fluctuation solar irradiation and variable load power for rural household load profile.

Contents

List of Figures

List of Tables

List of Abbreviations

PV	Photovoltaic energy
MPPT	Maximum Power Point Tracking
MPP	Maximum Power Point
DC	Direct Current
AC	Alternating Current
InCond	Incremental Conductance MPPT Technique
P&O	Perturb and Observe MPPT Technique
DOD	Depth of Discharge
BES	Battery Energy Storage
BESS	Battery Energy Storage Systems
SOC	State of Charge
AGM	Absorbent Glass Mat
SCs	Supercapacitors
ESR	Equivalent Series Resistance
HESS	Hybrid Energy Storage System
PWM	Pulse Width Modulation
IGBT	Insulated-Gate Bipolar Transistor
THD	Total Harmonic Distortion
BS-HESS	Battery-Supercapacitor Hybrid Energy Storage System
FFT	Fast Fourier Transform
FBC	Filtration-Based Controller
HPF	High-Pass Filter
LPF	Low-Pass Filter
SOC_{SC}	Supercapacitor State of Charge
MFs	Membership Functions

List of Symbols

Symbol	Meaning
E_{SC}, C	Energy stored in the SC, capacitance
Q, V	The stored charge (in Coulombs) and voltage (in Volt)
ε, ε_0	Dielectric constant and the permittivity of a vacuum
A, d	The thickness and the area between double layers of the capacitor
C_1, C_2	The equivalent capacitances in each electrical double layer
τ, R_{ESR}	Time constant and the equivalent series resistance of the SC
CdTe, a − Si	Cadmium telluride and amorphous silicon
$CuInSe_2$	Copper indium selenium
I, V	Output current and output voltage of PV cell
I_{pv}, V_{pv}	The terminal current and terminal voltage of PV array
I_{ph}, I_s	Light-generated current and PV saturation current
N_p, N_s	Number of parallel and series modules
R_s, R_{sh}	Series resistance and parallel resistance of PV cell
I_{sc}	Short-circuit current at STC (Standard Test Condition)
K_i, q	Short-circuit temperature coefficient and charge of electron
A, K	Ideality factor and Boltzmann's constant
E_g	Band-gap energy of semiconductor used in PV cell
T_{ref}, T	Reference temperature (25 °C) and actual temperature of PV cell
I_{rs}, G	Reverse saturation current at T_{ref} and solar irradiance
V_{oc}, N_{ser}	Open circuit voltage and number of series-connected PV cells
P_{PV}, f_s	Nominal power of the PV and switching frequency
C_a, C_1	PV array link capacitance and DC link capacitance
L_a, D	Boost converter inductor and diode
Dy, K_1	Duty cycle of the boost converter and constant of proportionality
V_{dc}, ΔV_0	Output voltage from boost converter and ripple of output voltage
ΔV_{PV}, ΔI_{La}	Change in PV voltage and ripple current of boost inductor
V_{mpp}, P_{pv}	PV array voltage at the MPP and PV array output power
f_{res}, R_d	Cut-off frequency and damping resistor
L_i, L_g	Inverter side inductance and grid side inductance
C_f, C_b	Filter capacity and system base capacitance
i, i^*	The battery current and the low-frequency current dynamics

it, Exp(s)	The battery extracted current and the extracted capacity
E_0, Q_b	The constant voltage and maximum battery capacity
Sel(s)	Represents the battery mode
P_{Batt}, P_{SC}	Battery power and SC power
P_{Load}	Power demand of the load
i_{self_dis}	Self-discharge current of SC
P_{LF}, P_{HF}	Power low-frequency components and power high-frequency components
dP	Mismatch power between PV power and load demand
I_{batt_peak}	Battery peak current
P_{batt_peak}	Battery peak power
SOC_{batt_avarge}	Average battery SOC
SOC_{batt_final}	Final battery SOC
I_{SC_peak}	SC peak current
P_{SC_peak}	SC peak power
SOC_{SC_final}	SC final SOC
A_i	Interfacial area between electrodes and electrolyte
C_m	Molar concentration
R_d, F	Molecular radius and Faraday constant
i_{SC}, V_{SC}	Supercapacitor current and voltage
C_T, R_{SC}	Total capacitance of the SC and total resistance of SC
N_e, N_A	Number of layers of electrodes and Avogadro constant
N_{pc}, N_{sc}	Number of parallel SCs and number of series SCs
α_1, α_2, and α_3	The rates of change of the SC voltage

Chapter 1
Introduction

1.1 Background

Nowadays, the most critical issue in the entire world is to meet the permanent growth of the energy demand. Some projections indicate that the global energy demand will almost triple by 2050 as in [1]. Moreover, the rapid depletion of the conventional power sources and their adverse impacts on the future of the planet has necessitated imperative researches for the renewable energy sources as alternative sources of energy. Also, the use of renewable energy sources is desired to improve energy efficiency which is essential to sustainable economic development. Furthermore, the use of renewable energy sources also reduces combustion of fossil fuels and consequent CO_2 emission which is the principal cause of greenhouse effect/global warming [2]. Among the renewable sources of energy, the PV energy and the wind energy have attracted great attention and can be considered as the most promising power technologies to generate the electricity. The PV energy and the wind energy are alternative to each other which will have the actual potential to be integrated with the electrical grid and satisfy the load dilemma to some degree. Also, the wind energy can be captured using large generators to generate great power capacity. Hence, the increased penetration of the wind energy generation systems is evident since it is clean, global, and having minimal operating cost requirements. On the other hand, the PV energy has shown great potential as another promising power technology to generate electricity since it is clean, global, and free and can be harnessed without emission of pollutants. In addition, the distributed PV systems, in contrast to the other renewable energy sources such as wind power generators, are more easily integrated into the electrical utility grids at any point. Therefore, the installation of PV systems has been growing rapidly in the last decades [3]. However, the PV energy and the wind energy are not entirely trustworthy, and they have some demerits such as their unpredictable nature and dependence on the environmental conditions such as the variations of the solar irradiance and the wind speed. Furthermore, the PV energy can be utilized only during the daylight

© Springer Nature Switzerland AG 2019
A. A. Elbaset et al., *Performance Analysis of Photovoltaic Systems with Energy Storage Systems*, https://doi.org/10.1007/978-3-030-20896-7_1

[4]. Both (if used independently) would have to be oversized to make them completely reliable, resulting in a higher total cost. Therefore, a merging of PV energy and wind energy into PV/wind hybrid generating system can attenuate their individual fluctuations, increase overall energy output, and generate more reliable power with higher quality to the electrical grid and the rural areas.

1.2   Photovoltaic Power Generation

Nowadays, the PV energy source has been one of the fastest growing renewable energy sources, which has annual growth rate around 55% over the last decade [2, 5]. The PV power generation utilizes the solar cells which convert the solar energy directly into electricity. At the heart of the PV systems is the PV cell, a semiconductor device which produces an electrical voltage and/or current when exposed to the sunlight [6]. Also, the PV cell generates a specified power according to its current-voltage (I-V) and power-voltage (P-V) characteristics. Therefore, the PV cells must be aggregated together to generate enough current and voltage for practical applications. In this regard, a PV module is formed by connecting several PV cells in series; the PV modules are connected in series to form a PV string to provide a greater output voltage. Then, the PV strings, in turn, are connected in parallel to form a PV array to increase the output current and generate enough power to be synchronized with the electrical grid. However, the incident solar irradiance on the PV array varies due to various reasons such as the variation of time in a day, the atmospheric effects such as clouds, and the latitude of the location. Therefore, the MPPT technique is implemented to regulate the output voltage and output current of PV array for extraction of the maximum power from the PV system during variation of the solar irradiance. Therefore, the PV system is equipped with a DC/DC boost converter to implement the MPPT technique and a three-phase voltage source inverter to be synchronized with the electrical grid [7].

1.2.1   Worldwide Annual Growth of PV Generation

The worldwide annual growth of the PV systems has the shape of an exponential curve during the period from 2007 to 2017, as illustrated in Fig. 1.1. For several years, the growth of PV systems was mainly driven by Japan and pioneering European countries. As a consequence, the cost of PV system installation has declined significantly due to the improvements in technology and design. Historically, the United States was the leader of installed PV systems for many years, and its total capacity amounted to 17 GW in 1996, more than any other country in the world at the time. Then, Japan was the world's leader in producing solar electricity until 2005, when Germany took the lead, and by 2016 it had a capacity of over 40 GW. However, in 2015, China became the world's largest producer of PV

power. China is expected to continue its rapid growth and to triple its PV capacity
to 70 GW by 2018 [8, 9]. Figure 1.1 shows the cumulative PV generation capacity
in the world. By the end of 2016, the global cumulative PV generation capacity has
increased from 30.3 GW in 2007 to roughly 306 GW, sufficient to supply between
1.3% and 1.8% of the global electricity demand. Then, the global cumulative PV
generation capacity has reached about 401.5 GW in 2017 [10]. Moreover, the global
cumulative PV generation capacity is projected to be more than 500 GW during the
period from 2017 to 2020. By the end of 2050, solar power is anticipated to become
the world's largest source of electricity, with concentrated solar power contributing
11%. Also, the global cumulative PV generation capacity is expected to grow to
4600 GW by the end of 2050 [11].

1.2.2 Photovoltaic Power Generation in Egypt

Egypt possesses an abundance of land, sunny weather, and high wind speeds, making it a prime location for utilization of the renewable energy sources. Egypt intends
to supply 20% of the electricity demand from the renewable energy sources by
2022, with wind providing 12%, hydropower 5.8%, and solar 2.2%. Egypt's Solar
Atlas states that Egypt is considered a "Sun Belt" country with 2000–3000 kWh/m²/
year of direct solar irradiation. In addition, the sun shines 9–11 hours a day from
north to south in Egypt with few cloudy days. Therefore, the solar energy plan aims
to install 3.5 GW by 2027, including 2.8 GW of PV and 700 MW of concentrated
solar power. Historically, the first solar thermal power plant was built at Kuraymat
in 2011; it has a total installed capacity of 140 MW. Also, a 10 MW power plant has
been operated in Siwa since March 2015, and the remaining plants are expected to
be implemented and operated in 2018. Recently, Egypt has embarked on an ambi-
tious project to build the biggest solar PV plant in the world at Benban, Aswan. The

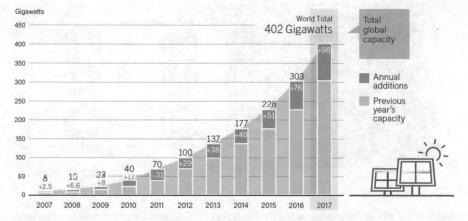

Fig. 1.1 Global PV generation capacity and annual additions, 2007–2017 [12]

PV station Benban project locates in the south of the Egyptian territory; the project has an estimated total cost of up to US$ 4 billion and will produce 1.8 GW of power when operational. The project site consists of a 37 km² plot divided into 39 projects of approximately 50 MW each [13].

1.3   Basics of Solar Cell

A PV system is made up of several PV solar cells, as illustrated in Fig. 1.2. An individual small PV cell can generate about 1 or 2 W of power approximately, depending on the type of material used. For higher power output, PV cells can be connected together to form higher power modules. In the market the maximum power capacity of the module is 1 kW; even though higher capacity is possible to manufacture, it will become cumbersome to handle more than 1 kW module. Depending upon the power plant capacity or based on the power generation, group of modules can be connected to form an array [14].

As per the statistics, the solar PV module world market is steadily growing at the rate of 30% per year. The reasons behind this growth are that the reliable production of electricity without fuel consumption anywhere there is light and the flexibility of PV systems [15]. Also, the solar PV systems using modular technology and the components of solar PV can be configured for varying capacity, ranging from watts to megawatts. Earlier, large variety of solar PV applications is found to be in industries, but now it is being used for commercial as well as for domestic needs.

One of the hindrance factors is the efficiency of the solar PV cell; in the commercial market, a cell efficiency of up to 18.3% is currently obtained, depending on the technology that is used. When it is related to the module efficiency, it is slightly lower than the cell efficiency. This is due to the blank spaces between the arrays of

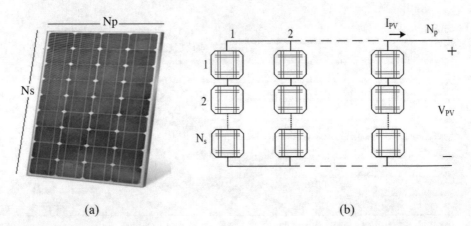

(a)   (b)

Fig. 1.2 PV module constructions and its circuit [14]. (**a**) Construction of PV module. (**b**) Circuit of PV module

solar cells in the module. The overall system efficiency includes the efficiency and the performance of the entire components in the system and also depends on the solar installation. Here there is another numerical drop in value when compared to the module efficiency, this being due to conductance losses, e.g., in cables. In the case of an inverter, it converts the DC output from the solar PV module to the AC grid voltage with a certain degree of efficiency. It depends upon conversion efficiency and the precision and quickness of the MPP tracking called tracking efficiency. The MPPT which is having an efficiency of 98–99% is available in the market; each and every MPPT is based on a particular tracking algorithm [16].

1.4   Types of PV with Storage Installations

Based on the electric energy production, PV modules can be arranged into arrays to increase electric output. Solar PV systems are generally classified based on their functional and operational requirements and their component configurations. There are three main types of solar PV and storage systems: grid-connected, PV-hybrid, and stand-alone solar PV system. They all have their advantages and disadvantages, and it really comes down to the customer's current energy supply and what they want to get out of the system. It can be classified into grid-connected and stand-alone systems [17].

1.4.1   Grid-Connected PV System

Grid-connected PV systems are usually installed to enhance the performance of the electric network by reducing the power losses and improving the voltage profile of the network. However, this is not always the case as these systems might impose several negative impacts on the network, especially if their penetration level is high [18]. A grid-tied system is a basic solar installation that uses a standard grid-tied inverter and does not have any battery storage, as illustrated in Fig. 1.3. This is perfect for customers who are already on the grid and want to add solar to their house. These systems can qualify for state and federal incentives which help to pay for the

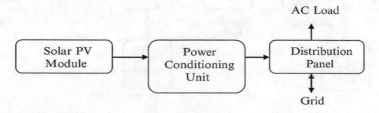

Fig. 1.3 Bock diagram of grid-connected PV system [14]

system. Grid-tied systems are simple to design and are very cost effective because they have relatively few components. The main objective of a grid-tied system is to lower your energy bill and benefit from solar incentives.

One disadvantage of this type of system is that when the power goes out, so does your system. This is for safety reasons because linemen working on the power lines need to know there is no source feeding the grid. Grid-tied inverters have to automatically disconnect when they don't sense the grid. This means that you cannot provide power during an outage or an emergency and you can't store energy for later use. You also can't control when you use the power from your system, such as during peak demand time.

But if a customer has a basic grid-tied system, they are not out of luck if they want to add storage later. The solution is doing an AC-coupled system where the original grid tied inverter is coupled with a battery back-up inverter. This is a great solution for customers who want to install solar now to take advantage of incentives but aren't ready to invest in the batteries just yet.

A customer can benefit from net metering because when the solar is producing more than they are using, they can send power back to the grid. But in times when the loads are higher than what the solar is producing, they can buy power from the utility. The customer is not reliant on the solar to power all his or her load. The main takeaway is that when the grid goes down, the solar is down as well and there's no battery backup in the system.

1.4.2   Grid-Tied System with Battery Backup

The next type of system is a grid-tied system with battery backup, otherwise known as a grid-hybrid system. As shown in Fig. 1.4, this type of system is ideal for customers who are already on the grid who know that they want to have battery backup. Good candidates for this type of system are customers who are prone to power outages in their area or generally just want to be prepared for outages.

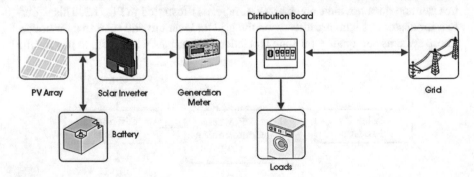

Fig. 1.4 Block diagram of grid-tied system with battery backup [19]

With this type of system, you get the best of both worlds because you're still connected to the grid and can qualify for state and federal incentives while also lowering your utility bill. At the same time, if there's a power outage, you have backup. Battery-based grid-tied systems provide power during an outage, and you can store energy for use in an emergency. You are able to back up essential loads such as lighting and appliances when the power is out. You can also use energy during peak demand times because you can store the energy in your battery bank for later use. Cons of this system are that they cost more than basic grid-tied systems and are less efficient. There are also more components. The addition of the batteries also requires a charge controller to protect them. There must also be a subpanel that contains the important loads that you want to be backed up. Not all the loads that the house uses on the grid are backed up with the system. Important loads that are needed when the grid power is down are isolated into a back-up subpanel [19].

1.4.3 Off-Grid System

Off-grid systems are great for customers who can't easily connect to the grid. This may be because of geographical location or high cost of bringing in the power supply. In most cases, it doesn't make much sense for a person connected to the grid to completely disconnect and do an off-grid system. The block diagram of stand-alone PV system with battery storage is shown in Fig. 1.5.

The benefits of an off-grid system are that a person can become energy self-sufficient and can power remote places away from the grid. You also have fixed energy costs and won't be getting a bill from your energy use. Another neat aspect of off grid system is that they are modular, and you can increase the capacity as your energy needs grow. You can start out with a small, budget-conscious system and add on over time.

Because the system is your only source of power, many off-grid systems contain multiple charging sources such as solar, wind, and generator. You have to consider weather and year-round conditions when designing the system. If your solar panels

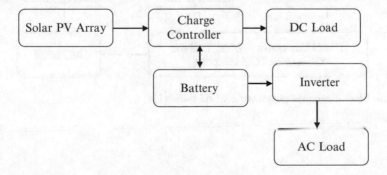

Fig. 1.5 Block diagram of stand-alone PV system with battery storage [14]

are covered in snow, you need to have another way to keep your batteries charged up [20]. You also will most likely want to have a back-up generator just in case your renewable sources are not enough at times to keep the batteries charged. One disadvantage is that off-grid systems may not qualify for some incentive programs. You must also design your system to cover 100% of your energy loads and hopefully even a little bit more. Off-grid systems have more components and are more expensive than a standard grid-tied system as well.

1.4.4   PV-Hybrid Systems

Hybrid systems generally refer to the combination of any two input sources; here solar PV can be integrated with diesel generator, wind turbines, biomass, or any other renewable on nonrenewable energy sources as illustrated in Fig. 1.6. Solar PV systems will generally use a battery bank to store energy output from the panels to accommodate a pre-defined period of insufficient sunshine; there may still be exceptional periods of poor weather when an alternative source is required to guarantee power production. PV-hybrid systems combine a PV module with another power sources – typically a diesel generator but occasionally another renewable supply such as a wind turbine. The PV generator would usually be sized to meet the base load demand, with the alternate supply being called into action only when essential. This arrangement offers all the benefits of PV in respect of low operation and maintenance costs but additionally ensures a secure supply [21].

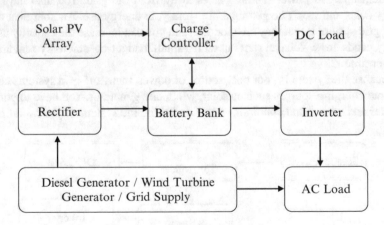

Fig. 1.6 Block diagram of photovoltaic hybrid system [14]

1.5   Energy Storage System

A fundamental characteristic of a PV system is that power is produced only while sunlight is available. For systems in which the PV is the sole generation source, storage is typically needed since an exact match between available sunlight and the load is limited to a few types of systems – for example, powering a cooling fan. In hybrid or grid-connected systems, where batteries are not inherently required, they may be beneficially included for load matching or power conditioning [22].

For off-grid and critical applications, storage systems are required; the most common medium of storage is the lead-acid battery. Presently researches are going on in the field of Li-ion batteries to implement the concept of fuel cells in solar PV systems. One of the most expensive components in the PV system is the battery. Under sizing the batteries will become more costly as the battery life cycle is significantly reduced at higher depth of discharge (DOD%). At a higher DOD, expected average number of charge-discharge cycles of a battery reduced. Further, a higher current discharge than the rating will dramatically reduce the battery life. This can be avoided by carefully sizing of the battery according to the "C-rating" during the system design. It signifies the maximum amount of current that can be safely withdrawn from the battery to provide adequate backup without causing any damage. A discharge more than the C-ratings may cause irreversible capacity loss due to the fact that the rate of chemical reactions taking place in the batteries cannot keep pace with the current being drawn from them [23]. The de-rating factor of the balance of system plays a significance role in boosting up the overall efficiency of the solar PV system.

1.6   Book Objectives

The main objectives of this book are summarized as follows:

1. Apply the MPPT technique for the PV system in order to extract the maximum power during variation of the environmental conditions.
2. Analyze the performance of the PV system with battery storage during variations of the solar irradiance in order to evaluate the effectiveness of the implemented MPPT techniques and the employed control strategy.
3. Addressing the performance analysis for the stand-alone PV system under the effect of the entry and exit sudden electrical loads.
4. Structure and modeling of stand-alone PV system with BS-HESS to reduce the dynamic stress and peak power demand of the battery by employing the appropriate control strategy.
5. Experimental work for performance analysis of stand-alone PV system with battery storage energy under variable solar irradiation and load profile.

1.7 Book Organization

To achieve the above objectives, the present book is organized in seven chapters in addition to a list of references. The chapters are organized as follows:

- *Chapter* 1: The main aim of this chapter is to present an introduction to the PV power generation worldwide and in Egypt. Also it presents an introduction to the principles for the solar cell. In addition, the types of photovoltaic power systems and energy storage systems have been reviewed.
- *Chapter* 2: This chapter introduces the energy storage systems in PV systems and discusses the classifications and types of batteries. Also, the focus was on lead-acid battery, and some properties of supercapacitors were reviewed. Finally, it gives an overview of previous works and methods based on the PV system and energy storage systems.
- *Chapter* 3: This chapter discusses the modeling of the fundamental elements in the off-grid PV systems. In addition, it introduces a simulation of two MPPT techniques that is implemented in the PV systems.
- *Chapter* 4: This chapter investigates a dynamic modeling, simulation, and control strategy of the stand-alone PV system with BES under variable load profile. Moreover, this chapter discusses the performance comparison of PV stand-alone system with BES in two cases of operation. In the first case, the system operates without and with BES under constant solar irradiation. In the second case, the PV system is connected to a BES and operates under a variable in solar irradiation.
- *Chapter* 5: This chapter proposes an optimal control strategy for a stand-alone PV system with BS-HESS. The proposed control strategy comprises of a low pass filter and fuzzy logic controller. The performance of the proposed system is compared to the conventional systems by Simulink with the setup of rural household load profile and the actual solar irradiation profile of a rainy day.
- *Chapter* 6: In this chapter, the experimental setup along with its components is implemented in renewable energy laboratory, faculty of industrial education, Suez University. This chapter includes two parts: the first part presents the experimental setup of an off-grid PV system, and the second part contains the experimental results and discussion.
- *Chapter* 7: This chapter reports the main conclusions from the book and summarizes the future research topics related to bookwork.

Chapter 2
Literature Survey

2.1 Introduction

A fundamental characteristic of a PV system is that power is produced only when sunlight is available. For systems in which the PV is the only generation source, storage is typically needed since an exact match between available sunlight and the load is limited to a few types of systems – for example, powering a cooling fan. In hybrid or grid-connected systems, where batteries energy storage (BESs) are not inherently required, they may be beneficially included for load matching or power conditioning. By far the most common type of storage is chemical storage, in the form of a battery energy storage (BES), although in some cases other forms of storage can be used [24]. For example, for small, short-term storage, a flywheel or capacitor can be used for storage, or for specific, single-purpose PV systems, such as water pumping or refrigeration, storage can be in the form of water or ice.

In any PV system that includes BESs, the BESs become a central component of the overall system which significantly affects the cost, maintenance requirements, reliability, and design of the PV system. Because of the large impact of BESs in a stand-alone PV system, understanding the properties of BESs is critical in understanding the operation of PV systems. The important BES parameters that affect the PV system operation and performance are the BES maintenance requirements, lifetime of the BES, available power, and efficiency. An ideal BES would be able to be charged and discharged indefinitely under arbitrary charging/discharging regimes and would have high efficiency, high energy density, low self-discharge, and below cost. These are controlled not only by the initial choice of the BES but also by how it is used in the system, particularly how it is charged and discharged and its temperature. However, in practice, no BES can achieve the above set of requirements, even if the normally dominant requirement for low cost is not considered. This chapter provides an overview of BES operation and uses for PV systems. The aim

© Springer Nature Switzerland AG 2019
A. A. Elbaset et al., *Performance Analysis of Photovoltaic Systems with Energy Storage Systems*, https://doi.org/10.1007/978-3-030-20896-7_2

of this chapter is to present the reader with enough information to understand how important it is to specify an appropriate type of BES and with sufficient capacity, for satisfactory use in a PV system.

2.2   Why Use a Battery Energy Storage in PV Systems?

The energy output from the solar PV systems is generally stored in BES deepening upon the requirements of the system. Mostly BESs are used in the stand-alone PV system, and in the case of grid-connected system, BESs are used as a back-up system [25]. The primary functions of the BES in a PV system are:

- It acts as a buffer store to eliminate the mismatch between power available from the PV array and power demand from the load. The power that a PV module or array produces at any time varies according to the amount of light falling on it (and is zero at nighttime). Most electrical loads need a constant amount of power to be delivered. The BES provides power when the PV array produces nothing at night or less than the electrical load requires during the daytime. It also absorbs excess power from the PV array when it is producing more power than the load requires [26].
- The BES provides a reserve of energy (system autonomy) that can be used during a few days of very cloudy weather or, in an emergency, if some part of the PV system fails.
- The BES prevents large, possibly damaging, voltage fluctuations. A PV array can deliver power at any point between a short circuit and an open circuit, depending on the characteristics of the load it is connected to.
- Supply surge currents: to supply surge or high peak operating currents to electrical loads or appliances [14].

2.3   BES Types and Classifications

Even BESs from the same manufacturers differ in their performance and its characteristics. Different manufacturers have variations in the details of their BES construction, but some common construction features can be described for almost all BESs. BESs are generally mass produced; it consists of several sequential and parallel processes to construct a complete BES unit. Different types of BESs are manufactured today, each with specific design for particular applications. Each BES type or design has its individual strengths and weaknesses. In solar PV system predominantly, lead-acid BESs are used due to their wide availability in many sizes, low cost, and well-determined performance characteristics. For low-temperature applications, nickel-cadmium cells are used, but their high initial cost limits their use in most PV systems. The selection of the suitable BES depends upon the application and the designer. In general, BESs can be divided into two major categories, primary and secondary BESs [21].

2.3.1   Primary BES

Primary BESs are non-rechargeable, but they can store and deliver electrical energy. Typical carbon-zinc and lithium BESs commonly used primary BESs. Primary BESs are not used in PV systems because they cannot be recharged.

2.3.2   Secondary BES

Secondary BESs are rechargeable, and they can store and deliver electrical energy. Common lead-acid BESs used in automobiles and PV systems are secondary BESs. The BESs can be selected based on their design and performance characteristics. The PV designer should consider the advantages and disadvantages of the BESs based on its characteristics and with respect to the requirements of an application. Some of the important parameters to be considered for the selection of BES are lifetime, deep-cycle performance, tolerance to high temperatures and overcharge, maintenance, and many others [27]. Examples of rechargeable BES systems are:

- Lead-acid
- Nickel-cadmium (Ni-Cd)
- Nickel-iron
- Nickel-metal hydride (Ni-MH)
- Rechargeable lithium of various types, especially lithium-ion

This book's main focus is on the use of lead-acid BESs in stand-alone PV systems.

2.4   Battery Energy Storage Characteristics

The use of BESs in PV systems differs from the use of BESs in other common BES applications. For PV systems, the key technical considerations are that the BES experiences a long lifetime under nearly full discharge conditions. Common rechargeable BES applications do not experience both deep cycling and being left at low states of charge for extended periods of time. For example, in BESs for starting cars or other engines, the BES experiences a large, short current drain but is at full charge for most of its life. Similarly, BESs in uninterruptible power supplies are kept at full charge for most of their life. For BESs in consumer electronics, the weight or size is often the most important consideration [28]. This section provides an overview of the critical BES characteristics or specifications, including BES voltage, capacity, charging/discharging regimes, efficiency, etc.

2.4.1 Battery Energy Storage Charging

In a stand-alone PV system, the ways in which a BES is charged are generally much different from the charging methods BES manufacturers use to rate BES performance. The BES charging in PV systems consists of three modes of BES charging: normal or bulk charge, finishing or float charge, and equalizing charge [29].

- *Bulk or Normal Charge*: It is the initial portion of a charging cycle performed at any charge rate, and it occurs between 80% and 90% SOC. This will not allow the cell voltage to exceed the gassing voltage.
- *Float or Finishing Charge*: It is usually conducted at low to medium charge rates. When the BES is fully charged, most of the active material in the BES has been converted to its original form, generally voltage/current regulation that is required to limit the overcharge supplied to the BES.
- *Equalizing Charge*: It consists of a current-limited charge to higher voltage limits than set for the finishing or float charge. It is done periodically to maintain consistency among individual cells. An equalizing charge is typically maintained until the cell voltages, and specific gravities remain consistent for a few hours.

2.4.2 Battery Energy Storage Discharging

- *Depth of Discharge (DOD):* The DOD of BES is defined as the percentage of capacity that has been withdrawn from a BES compared to the total fully charged capacity. The two common qualifiers for DOD in PV systems are the allowable or maximum DOD and the average daily DOD.
- *Allowable DOD*: The maximum percentage of full-rated capacity that can be withdrawn from a BES is known as its allowable DOD. In stand-alone PV systems, the low voltage load disconnect set point of the BES charge controller dictates the allowable DOD limit at a given discharge rate. Depending on the type of BES used in a PV system, the design allowable DOD may be as high as 80% for deep-cycle, motive power BESs, to as low as 15–25%. The allowable DOD is related to the autonomy, in terms of the capacity required to operate the system loads for a given number of days without energy from the PV array.
- *Average Daily DOD*: This is the percentage of the full-rated capacity that is withdrawn from a BES with the average daily load profile. If the load varies seasonally, the average daily DOD will be greater in the winter months due to the longer nightly load operation period. For PV systems with a constant daily load, the average daily DOD is generally greater in the winter due to lower BES temperature and lower rated capacity. Depending on the rated capacity and the average daily load energy, the average daily DOD may vary between only a few percent in systems designed with a lot of autonomy or as high as 50% for marginally sized BES systems. The average daily DOD is inversely related to auton-

omy; meaning that systems designed for longer autonomy periods (more capacity) have a lower average daily DOD.

- *State of Charge (SOC):* This is defined as the amount of energy as a percentage of the energy stored in a fully charged BES. Discharging a BES results in a decrease in SOC, while charging results in an increase in SOC.
- *Autonomy:* Generally, autonomy refers to the time a fully charged BES can supply energy to the system loads when there is no energy supplied by the PV array. Longer autonomy periods generally result in a lower average daily DOD and lower the probability that the allowable (maximum) DOD or minimum load voltage is reached.
- *Self-Discharge Rate*: In open-circuit mode without any charge or discharge current, a BES undergoes a reduction in SOC, due to internal mechanisms and losses within the BES. Different BES types have different self-discharge rates, the most significant factor being the active materials and grid alloying elements used in the design.
- *Battery Lifetime:* Battery lifetime is dependent upon a number of design and operational factors, including the components and materials of BES construction, temperature, frequency, DODs, and average SOC and charging methods.
- *Temperature Effects:* For an electrochemical cell such as a BES, temperature has important effects on performance. As the temperature increases by 10° C, the rate of an electrochemical reaction doubles, resulting in statements from BES manufacturers that BES life decreases by a factor of two for every 10° C increase in average operating temperature. Higher operating temperatures accelerate corrosion of the positive plate grids, resulting in greater gassing and electrolyte loss. Lower operating temperatures generally increase BES life. However, the capacity is reduced significantly at lower temperatures, particularly for lead-acid BESs. When severe temperature variations from room temperatures exist, BESs are located in an insulated or other temperature-regulated enclosure to minimize BES.
- *Effects of Discharge Rates:* The higher the discharge rate or current, the lower the capacity that can be withdrawn from a BES to a specific allowable DOD or cutoff voltage. Higher discharge rates also result in the voltage under load to be lower than with lower discharge rates, sometimes affecting the selection of the low voltage load disconnect set point. At the same BES voltage, the lower the discharge rates, the lower the BES SOC compared to higher discharge rates.
- *Corrosion*: The electrochemical activity resulting from the immersion of two dissimilar metals in an electrolyte or the direct contact of two dissimilar metals causing one material to undergo oxidation or lose electrons and causing the other material to undergo reduction or gain electrons. Corrosion of the grids supporting the active material in a BES is an ongoing process and may ultimately dictate the BES's useful lifetime. BES terminals may also experience corrosion due to the action of electrolyte gassing from the BES and generally require periodic cleaning and tightening in flooded lead-acid types [14, 30].

Table 2.1 Compare the properties for some types of BESs [31]

Specifications	Lead-acid	Ni-Cd	Ni-MH	Li-ion		
				Cobalt	Manganese	Phosphate
Specific energy density (Wh/Kg)	30–50	45–80	60–120	150–190	100–135	90–120
Internal resistance (mΩ)	<100 12 V peak	100–200 6 V peak	200–300 6 V peak	150–300 7.2 V	25–75 per cell	25–50 per cell
Cycle life (80% discharge)	200–300	1000	300–500	500–1000	500–1000	1000–2000
Fast-charge time	8–16 h	1 h typical	2–4 h	2–4 h	1 h or less	1 h or less
Overcharge tolerance	High	Moderate	Low	Low. Can't tolerate trickle charge		
Self-discharge/ month	5%	20%	30%	<10%		
Cell voltage	2 V	1.2 V	1.2 V	3.6 V	3.8 V	3.3 V
Charge Cutoff Voltage (V/cell)	2.4	Full charge detection by voltage signature		4.2	4.2	3.6
Discharge cutoff voltage (V/cell)	1.75	1	1	2.5–3	2.5–3	2.8
Peak load current Best result	5A 0.2A	20A 1A	5A 0.5A	>3A <1A	>30A <10A	>30A <10A
Charge temperature	−20 to 50 °C	0 to 45 °C		0 to 45 °C		
Discharge temperature	−20 to 50 °C	−20 to 65 °C		−20 to 60 °C		
Maintenance requirement	3–6 months	30–60 days	60–90 days	Not required		
In use since	Late 1800s	1950	1990	1991	1996	1999

2.4.3 Compare the Characteristics of Some Types of BESs

There are a large number of BES parameters. Depending on which application the BES is used for, some parameters are more important than others. The following is a list of parameters that may be specified by a manufacturer for a given type of BES which is listed in Table 2.1. For example, in a typical BES for a general car, the energy density is not relevant – a BES is a small fraction of the total BES weight, and consequently, this parameter would typically not be listed for a conventional car BES. However, in electric vehicle applications, the BES weight is a significant fraction of the overall weight of the vehicle, and so the energy densities will be given [26].

2.5   Lead-Acid Battery Energy Storage

Lead-acid BESs are the most commonly used type of BES in PV systems. Although lead-acid BESs have a low energy density, only moderate efficiency, and high maintenance requirements, they also have a long lifetime and low costs compared to other BES types. One of the singular advantages of lead-acid BESs is that they are the most commonly used form of BES for most rechargeable BES applications (e.g., in starting car engines) and therefore have a well-established, mature technology base [32].

A lead-acid BES or cell in the charged state has positive plates with lead dioxide (PbO_2) as an active material, negative plates with high surface area (spongy) lead as an active material, and an electrolyte of the sulfuric acid solution in water (about 400–480 g/mL, density 1.241.28 kg/L). On discharge, the lead dioxide of the positive plate and the spongy lead of the negative plate are both converted to lead sulfate in Fig. 2.1. Lead-acid BESs store energy by the reversible chemical reaction shown below [33].

- Lead-acid overall reaction [32]:

$$\begin{array}{cc} \text{Charged} & \text{Discharged} \\ PbO_2 + Pb + 2H_2SO_4 & \Leftrightarrow \quad 2PbSO_4 + 2H_2O \end{array} \tag{2.1}$$

- Lead-acid positive terminal reaction:

$$\begin{array}{cc} \text{Charged} & \text{Discharged} \\ PbO_2 + 3H^+ + HSO_4^- + 2e^- & \Leftrightarrow \quad PbSO_4 + 2H_2O \end{array} \tag{2.2}$$

- Lead-acid negative terminal reaction:

$$\begin{array}{cc} \text{Charged} & \text{Discharged} \\ Pb + HSO_4^- & \Leftrightarrow \quad PbSO_4 + H^+ + 2e^- \end{array} \tag{2.3}$$

Note that the electrolyte (sulfuric acid) takes part in this basic charge and discharge reactions, being consumed during discharge and regenerated during charge. This means that the acid concentration (or density) will change between charge and discharge. It also means that an adequate supply of acid is needed at both plates when the BES is discharging in order to obtain the full capacity.

The lead-acid BES system has a nominal voltage of 2.0 V/cell as shown in Fig. 2.2. The typical end voltage for discharge in PV systems is 1.8 V/cell, and the typical end voltage for charging in PV systems varies between 2.3 and 2.5 V/cell, depending on the BES, controller, and system type. The relation of open circuit voltage to SOC is variable but somewhat proportional. However, if charging or discharging is interrupted to measure the open circuit voltage, it can take a long time (many hours) for the BES voltage to stabilize enough to give a meaningful value.

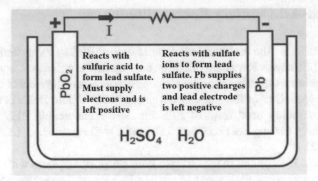

Fig. 2.1 Chemical reaction when a battery is being discharged [33]

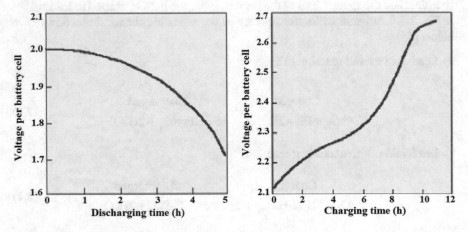

Fig. 2.2 Charge and discharge characteristic of lead-acid BES voltage per cell [31]

2.6   Calculating Battery Size for a PV System

We now list the full process of correctly calculating the capacity required for a particular battery type in a specific PV system [34].

2.6.1   Select the Appropriate Voltage

This is defined by the load (and PV array) nominal voltage unless some DC/DC converter is present in the system. This sets the number of cells or blocks that must be connected in series.

2.6.2   Define Maximum Depths of Discharge

These must be defined for each battery type according to the mode of operation [26].

- The maximum DOD for autonomy reserve is normally set at 80% for a lead-acid battery.
- The maximum daily DOD may either be set arbitrarily (e.g., a figure of 20–30% is common).
- For seasonal storage (if used), a maximum DOD needs to be set.
- For open batteries in most PV systems, a charge rate faster than the 10-hour rate is not recommended.
- For sealed batteries, another consideration is the highest overcharge current that can be sustained with efficient gas recombination, and this is temperature-dependent.

2.6.3   Calculate the Battery Capacity

The battery type recommended for using in solar PV system is deep-cycle battery. Deep-cycle battery is specifically designed to be discharged to low energy level and rapid recharged or cycle charged and discharged day after day for years. The battery should be large enough to store sufficient energy to operate the appliances at night and cloudy days. To find out the size of the battery, calculate as follows [35]:

(a) Calculate total Watt-hours per day used by loads.
(b) Divide the total Watt-hours per day used by 0.85 for battery loss.
(c) Divide the answer obtained in item (b) by 0.8 for DOD.
(d) Divide the answer obtained in item (c) by the nominal battery voltage.
(e) Multiply the answer obtained in item (d) with days of autonomy (the number of days that you need the system to operate when there is no power produced by PV panels) to get the required Ampere-hour capacity of deep-cycle battery.

$$\text{Battery Capacity}(\text{Ah}) = \frac{\text{Total Wh per day used by loads}}{0.85 \times 0.8 \times \text{normal battery voltage}} \times \text{day of autonomy} \quad (2.4)$$

2.7   The Supercapacitor Energy Storage System in PV System

Supercapacitors (SCs) are based on electrochemical cells that contain two conductor electrodes, an electrolyte and a porous membrane that permits the transit of ions between the two electrodes. Thus, the presented layout is similar to the electrochemical cells of batteries. The main difference between SC (or ultracapacitors or double-layer capacitors) and batteries lies in the fact that no chemical reactions occur in the cells but the energy is stored electrostatically in the cell [36].

In SCs, the electrodes and the electrolyte are electrically charged (the cathode is positively charged, the anode is negatively charged, and the electrolyte contains both positive and negative ions) as shown in Fig. 2.3. At each of the electrode surfaces, there is an area that interfaces with the electrolyte, and it is in each of these areas where the phenomenon of the "electrical double layer" occurs. By applying a voltage between the electrodes, both the electrodes and the electrolyte become polarized. This means that the positive charge of the cathode is transferred to the area interfacing with the electrolyte, forming a layer of positive ions. In turn, the negative ions of the electrolyte are transferred to the same electrolyte/cathode interface, forming a negative charge balancing layer of ions. These two layers build up an "electrical double layer." The mechanism behind the operating principle of such a double layer can be explained using the Helmholtz model [37].

The model establishes that the two layers are separated by a layer of solvent molecules of the electrolyte, called the inner Helmholtz plane. This layer of solvent molecules actually separates the positive and negative charges of the electrode and electrolyte, thus acting as a dielectric. Ultimately, there is a potential difference between the two layers of positive and negative ions derived from the electric field within them, and the double layer can be taken to resemble a capacitor (the described double layer concept can be observed in Fig. 2.3; see also Fig. 2.4).

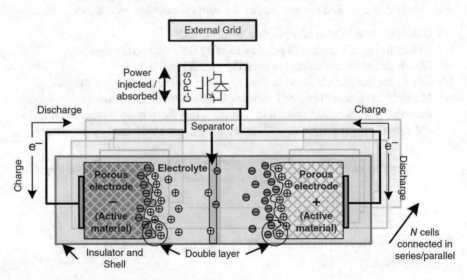

Fig. 2.3 The illustrative topology of a SC, depicting the electrical double layers at each electrode/electrolyte interface [40]

Fig. 2.4 Supercapacitor
modules from Maxwell
Technologies, Appendix A

Therefore, the magnitude of the electrical potential (in Volts) between the two layers of positive and negative ions at each electrode/electrolyte interface, in conjunction with the resultant capacitance (in Farads), determines the energy stored in the SC. Thus [38],

$$E_{SC} \left(\text{joules} \right) = \frac{1}{2} C V^2 \tag{2.5}$$

The voltage generated in the cell is dependent on the strength of the electric field between the layers building up each of the "electrical double layers" described above. This electric field is, in turn, proportional to the amounts of positive and negative ions located at the electrode/electrolyte interface. So to avoid the transfer of ions between the two layers of positive and negative ions, thus decreasing the voltage within the double layers, the breakdown voltage of the dielectric should be maximized. As noted before, this dielectric is provided by solvent molecules of the electrolyte. In this way, the selection of the electrolyte is key to ensuring the maximum energy capacity. Usually, both aqueous and organic electrolytes are commonly found, the latter being the most common type. With aqueous electrolytes, a cell voltage of around 1 V can be obtained, while it can be increased up to 2.5 V by using organic types.

As stated in Eq. (2.6), the second factor affecting the energy capacity of SCs is the capacitance of the cell. The capacitance (in Farads) of a capacitor is given by the quotient between the stored charge (in Coulombs) per unit of voltage (in Volts), so

$$C = Q / V \tag{2.6}$$

In addition, the capacitance can be expressed as a function of the permeability of the dielectric, its thickness, and the area holding each of the layers of the electrical double layer. Then,

$$C = \varepsilon \, \varepsilon_0 \, \frac{A}{d} \qquad\qquad (2.7)$$

where ε is the dielectric constant, ε_0 is the permittivity of a vacuum, A is the effective area of the surface of the electrode, and d is the dielectric thickness.

In order to maximize the capacitance, different metal-oxide electrodes, electronically conducting polymer electrodes, and activated carbon electrodes are used in industry. These materials are porous, so they can maximize the effective area of the electrode in which ions can be allocated. The most common types are the ones based on activated carbon since they can lead to SCs with a high energy density and capacitances around 5000 F [36], that is, capacities up to 1000 times per unit volume more than those of conventional electrolytic capacitors.

About the distribution of capacitance between the two electrical double layers in the cell, we can distinguish between symmetrical and unsymmetrical SCs. Symmetrical ones are those with the same effective area in both electrodes. Since the cell can be considered to resemble two capacitors in series (given by the two double layers at each electrolyte/electrode interface), the total capacitance can be formulated as

$$C_{eq} = \frac{C_1 C_2}{C_1 + C_2} \qquad\qquad (2.8)$$

where C_1 and C_2 are the equivalent capacitances in each electrical double layer. As mentioned, the electrolyte and electrode materials have a fundamental influence on the energy and power capacity of the SC and also on its dynamic behavior. To be precise and with reference to the SC dynamics, one defining parameter is the

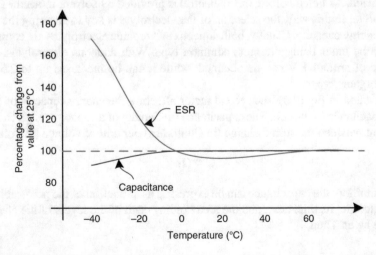

Fig. 2.5 The capacitance and the ESR as temperature-dependent characteristics. Appendix A

so-called charge/discharge time constant, τ. This is given by the product of the equivalent series resistance (ESR) of the SC and its capacitance. Thus,

$$\tau = R_{ESR}C \tag{2.9}$$

The time constant is the time needed to discharge 63.2% of full capacity with a current limited only by the internal resistance – or the ESR as it is commonly known – of the SC. The ESR weights the losses in the SC while charging and discharging, that is, those associated with the movement of ions within the electrolyte and across the separator. The ESR is normally in the range of milliohms and is a temperature-dependent parameter, as presented in Fig. 2.5.

Apart from the ESR and the capacitance, the third characteristic parameter for the SC is the leakage resistance, which weights the self-discharge of the cell. This resistance is much higher than the ESR. All three parameters – the capacitance, the ESR, and the leakage resistance – can be found in manufacturers' datasheets, and from them, averaged models for SCs can be built.

SCs are characterized by offering high ramp power rates, high cyclability, high round-trip efficiency (of up to 80%), and a high specific power, in W/kg, and power density, in W/m^3 (10 times more than for conventional batteries). The latter characteristic defines SCs as well suited for applications that impose major volumetric restrictions. On the other hand, major drawbacks of the technology are related to its high self-discharge rates (of up to 20% of the rated capacity in only 12 h) and its limited applicability to situations where high power and energy are needed. In fact, the development of SCs is mostly focused in fields such as automotive and portable devices. Finally, it is worth noting that as a short timescale, SCs are unsuitable in that they are expensive in comparison with other competitors such as flywheels. Their cost is estimated at 10 times the cost per kWh of flywheels.

SCs are, in general, young technologies. The first prototypes were developed in 1957 by H.I. Becker (General Electric). However, the first related studies were carried out in the nineteenth century by Helmholtz, who discussed the electrical behavior of a metal surface while immersed in an electrolyte. Currently, intense research activity is underway to scale up SC size and to improve their performance, so that they will be suitable for both stationary and nonstationary applications – such as in the fields of electromobility and PV system.

2.8 Literature Survey of Previous Works

This section provides literature survey of previous works and methods about a stand-alone PV system with the battery-supercapacitor hybrid energy storage system (HESS). The block diagram of the system is shown in Fig. 2.6. There are many researches on the modeling of PV array, MPPT, and half-bridge bidirectional DC/DC converter, and energy storage systems have been studied.

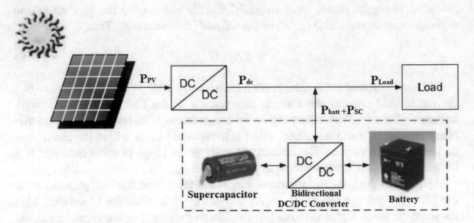

Fig. 2.6 Block diagram of stand-alone PV system with BS-HESS

2.8.1   Review of Related Researches About PV Modeling

K. Ishaque, Z. Salam, and H. Taheri [39] presented an improved modeling approach for the two-diode model of PV module based on four parameters. The proposed model is tested on six PV modules of different types (multi-crystalline, monocrystalline, and thin film) from various manufacturers.

J. Maherchandani, Ch. Agarwal, and M. Sahi [40] presented an efficient and accurate single-diode model for the estimation of the solar cell parameters using the hybrid genetic algorithm and Nelder-Mead simplex search method from the given voltage-current data.

Z. Ahmad and S. Singh [41] presented a method to extract the internal parameters such as ideality factor, series, and shunt resistance of any solar PV cell using block diagram modeling of PV cell/module using Matlab/Simulink model.

S. Lineykin, M. Averbukh, and A. Kuperman [42] implemented the single-diode equivalent circuit to modeling amorphous silicon PV modules. This approach combined numerical solution of two transcendental and two regular algebraic equation systems with single parameter fitting procedure.

B. Chitti Babu and Suresh Gurjar [43] presented modeling approach of PV modules using an ideal two-diode model. This model was simplified by omitting series and shunt resistances, only four unknown parameters from the datasheet were required to analyze the proposed model.

2.8.2   Review of Related Researches About MPPT of PV System

MPPT is a power electronic device with computer that is connected between PV power source and load to extract maximum power from a PV module and satisfy the highest efficiency. Until now numerous of MPPT techniques have been developed to increase the efficiency of the PV system and satisfy the optimal MPPT. These techniques vary in various aspects such as tracking speed, oscillations around MPP, cost, and hardware required for implementation. Most famous MPPT controllers available are fractional open circuit voltage, fractional short circuit current [44], hill climbing [45, 46], P&O [48, 49], InCond [50, 51], incremental resistance [50], ripple correlation control [51], fuzzy logic [52], artificial neural networks [53], particle swarm optimization [54], and sliding mode [55].

P. Francisco and M. Ordonez [56] presented the zero-oscillation, adaptive-step P&O MPPT strategy for solar PV panels. This combined strategy reduced steady-state losses and improved transient behavior during slope changes irradiance while maintaining a similar implementation complexity.

M. Killi and S. Samanta [57] suggested the positive sign of current change to avoid the problem, but this solution is only for increasing of irradiance and lacking information about rapid decreasing of weather.

2.8.3   Review of Related Researches About Half-Bridge Bidirectional DC/DC Converter

To appropriately interface the batteries and the SCs in the HESS (such as in hybrid electric vehicle), a bidirectional DC/DC converter is required to control the power flow in two directions. R M. Schupbach and J.C. Balda [58] presented analysis, design, and comparative study of several bidirectional non-isolated DC-DC converter topologies.

F. A. Himmelstoss and M. E. Ecker [59] who introduced a bidirectional DC/DC half-bridge converter presented analyses with a view to obtaining maximum voltage and current ratings for the elements, rms values for the semiconductor devices, and a rough approximation of the losses.

J. Cao and A. Emadi [60] presented compared to the conventional HESS design, which uses a larger DC/DC converter to interface between the ultracapacitor and the battery/dc link to satisfy the real-time peak power demands and design a much smaller DC/DC converter working as a controlled energy pump to maintain the voltage of the ultracapacitor at a value higher than the battery voltage for the most city driving conditions.

2.8.4   Review of Related Researches About a Stand-Alone PV System with HESS

In this paragraph review of previous research on the PV stand-alone systems is related with a BS-HESS. W. Jing, C. H. Lai, M. L. Dennis Wong, and W. S. H. Wong [61] illustrate in islanded microgrid system the battery tenders to be the most vulnerable element in terms of durability. Poorly managed battery charge/discharge process is one of the main life-limiting factors. To improve the battery life, a novel energy storage system topology and a power allocation strategy are proposed.

I. Shchur and Y. Biletskyi [62] discussed the stand-alone PV system; under variable of daily and weather solar irradiation, special devices are used for energy storage. In order to remove stress from batteries during sudden load change, it is advisable to use HESS by adding a *SC* module to battery.

J. Cao and A. Emadi [60] presented in their paper a battery/ultracapacitor hybrid energy storage system for electric, hybrid, and plug-in hybrid electric vehicles. Compared to the conventional HESS design, which uses a larger DC/DC converter to interface between the ultracapacitor and the battery/dc link to satisfy the real-time peak power demands, the proposed design uses a much smaller DC/DC converter working as a controlled energy pump to maintain the voltage of the ultracapacitor at a value higher than the battery voltage for the most city driving conditions.

Lee Wai Chong, Yee Wan Wong, Rajprasad Kumar Rajkumar, and Dino Isa [63] presented comparison between the stand-alone PV system with BS-HESS and the conventional stand-alone PV system with battery-only storage system for a rural household. Stand-alone PV system with passive BS-HESS and semi-active BS-HESS is presented in this study.

2.9   Summary

This chapter presented and reviewed the importance of energy storage systems in PV system. The energy from PV systems is generally stored in BES deepening upon the requirements of the system. The main functions of the BES in a PV system are used to store energy to eliminate the mismatch between PV power and power load demand. Most electrical loads need a constant amount of power to be delivered. The BES provides power when the PV array produces nothing at night or less than the electrical load requires during the daytime. The BES provides a reserve of energy that can be used during a few days of very cloudy weather or in an emergency. The BES prevents large, possibly damaging, voltage fluctuations. The types of BES used with PV systems were also clarified, and a comparison was made between these types in terms of charging and discharging characteristics. This chapter also explained the details of the most common type of batteries used in PV systems, the lead-acid battery. The operation of lead-acid battery, internal structure, and common types has been explained. In addition, this chapter presents a review of SC in terms of working theory, internal components, and general properties.

Chapter 3
Modeling of Maximum Power Point Tracking for Stand-Alone PV Systems

3.1 Introduction

The PV systems utilize semiconductor materials and electronic technology to convert the incident sunlight into electricity. At the heart of the PV system is the PV cell, a semiconductor material which generates electrical voltage and/or current when exposed to the solar irradiance. The PV cells generate electricity via the PV effect, in which semiconductor holes and electrons freed by photons from the incident solar irradiance are dragged to opposite terminals of the PV cell by the resulting electric field [6]. The PV cell generates a specified power according to its current-voltage (I-V) and power-voltage (P-V) characteristics. Thus, the PV cells must be aggregated together to produce enough current and voltage for practical applications. In this regard, a PV module is formed by connecting several PV cells in series; the PV modules are connected in series to form a PV string. The PV strings, in turn, are connected in parallel to form a PV array in order to generate adequate voltage and power to be integrated with the electrical grid.

The incident solar irradiance on the PV array varies due to various reasons such as the variation of time in a day, the atmospheric effects such as clouds, and the latitude of the location. Therefore, the MPPT techniques are implemented to regulate the output voltage and current of PV array for extraction of the maximum power during variation of the solar irradiation. In addition, the PV systems are equipped with a DC/DC converter to implement the MPPT technique [64]. This chapter discusses the fundamental components of the stand-alone PV systems. Moreover, this chapter introduces the study and design of two techniques of MPPT that implemented in the PV conversion systems, namely, the perturb and observe (P&O) MPPT technique and the incremental conductance (InCond) MPPT technique.

© Springer Nature Switzerland AG 2019

A. A. Elbaset et al., *Performance Analysis of Photovoltaic Systems with Energy Storage Systems*, https://doi.org/10.1007/978-3-030-20896-7_3

3.2   Principle of PV Conversion Systems

Without pollution or greenhouse gas emission, a PV conversion system converts the sunlight directly into electricity. The basic element of a PV conversion system is the PV cell. The PV cell is basically made up of a semiconductor material (P-N junction) that able to generate the electric current when being exposed to the sunlight irradiation. Figure 3.1 illustrates the photocurrent generation principle of the PV cell. These PV modules can be grouped in series and/or parallel to form a PV array as depicted in Fig. 3.2. The PV modules are connected in parallel to increase the output current and connected in series to provide a greater output voltage.

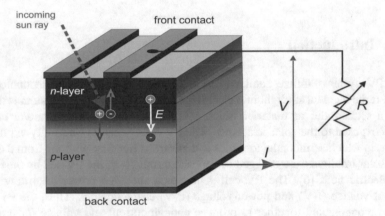

Fig. 3.1 Photocurrent generation principle of the PV cell [67]

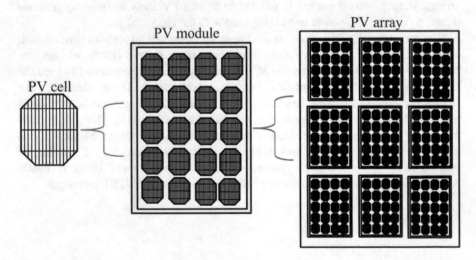

Fig. 3.2 PV cell, PV module, and PV array [67]

The dynamic performance of the PV conversion systems depends on the design quality of the PV cells and the operating conditions. The major families of PV cells include monocrystalline technology, polycrystalline technology, and thin-film technology. The monocrystalline and polycrystalline technologies are based on micro-electronic manufacturing technology and their efficiency generally between 9% and 12% for polycrystalline and between 10% and 15% for monocrystalline. For the thin-film technology, the efficiency for *CdTe* is 9%, 10% for $a - Si$, and 12% for *CuInSe*$_2$ [65]. Therefore, the monocrystalline PV cells are the most employed in the PV systems since they have the highest efficiency. Moreover, the electrical characteristics of PV array such as the output voltage, current, and power vary according to the changes of the environmental conditions such as the solar irradiance and the temperature. Therefore, the effect of the environmental condition's variations must be considered in the design of PV array so that any change in the solar irradiance or temperature should not adversely affect the output power of PV array [66].

3.3 The Main Components of Stand-Alone PV Systems

The Simulink block diagram of a stand-alone PV system is manifested in Fig. 3.3. The fundamental element of a stand-alone PV system is the PV array, which converts the solar energy directly into the electric energy. Then, the DC/DC boost converter is employed to step up the output voltage from PV array to be compatible with the electrical loads. Moreover, the MPPT technique is implemented on the boost converter to extract the maximum power from the PV system during variation of the solar irradiance. In the following, the basic elements that employed in stand-alone PV systems are discussed in detail.

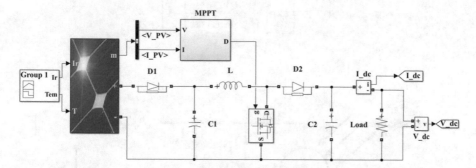

Fig. 3.3 Simulink block diagram of the stand-alone PV system with DC/DC converter

3.3.1 The Equivalent Circuit of the PV Model

The basic element of the PV conversion system is the PV cell, which is a just simple P-N junction. The equivalent circuit of the PV cell based on the well-known single-diode model is shown in Fig. 3.4. It includes the current source (photocurrent), a diode (D), series resistance (R_s) that describes the internal resistance to flow of current, and parallel resistance (R_{sh}) that represents the leakage current. The current-voltage (I-V) characteristics of the PV cell can be expressed as follows [1, 66]:

$$I = I_{ph} - I_s \left\{ \exp\left[\frac{q(V+IR_s)}{A \cdot K \cdot T}\right] - 1 \right\} - \left(\frac{V+IR_s}{R_{sh}}\right) \tag{3.1}$$

The light-generated current (I_{ph}) mainly depends on the solar irradiance and working temperature of PV cell, which is expressed as follows:

$$I_{ph} = \left[I_{sc} + K_i\left(T - T_{ref}\right)\right] \cdot \left(\frac{G}{1000}\right) \tag{3.2}$$

The PV saturation current (I_s) varies as a cubic function of the PV cell temperature (T), and it can be described as follows:

$$I_s = I_{rs}\left(\frac{T}{T_{ref}}\right)^3 \exp\left[\frac{q \cdot E_g}{K \cdot A} \cdot \left(\frac{1}{T_{ref}} - \frac{1}{T}\right)\right] \tag{3.3}$$

The reverse saturation current (I_{rs}) can be approximately obtained as follows:

$$I_{rs} = \frac{I_{sc}}{\left[\exp\left(\frac{qV_{oc}}{N_{ser} \cdot K \cdot A \cdot T}\right) - 1\right]} \tag{3.4}$$

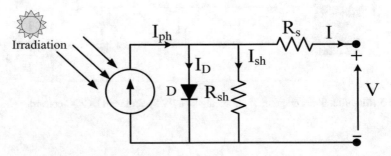

Fig. 3.4 Equivalent circuit of the PV cell

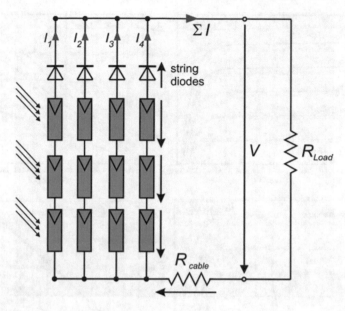

Fig. 3.5 Configuration of the PV array

The PV cell generates a specified power according to its current-voltage (I-V) and power-voltage (P-V) characteristics. Therefore, the PV cells must be aggregated together to generate sufficient current and voltage for practical applications. In this regard, a PV module is formed by connecting several PV cells in series; the PV modules are connected in series to form a PV string to provide a greater output voltage. Then, the PV strings, in turn, are connected in parallel to form a PV array to increase the output current and generate sufficient power to be synchronized with the electrical grid, as illustrated in Fig. 3.5. The electrical characteristics of PV array such as output voltage, output current, and output power can be simulated with regard to the variations of the environmental conditions such as the solar irradiance and temperature. Figure 3.6 illustrates the current-voltage (I-V) and the power-voltage (P-V) characteristics of a typical PV array during variation of the solar irradiance and temperature. As illustrated in Fig. 3.6a, the solar irradiance has a great effect on the short-circuit current (I_{sc}), while in Fig. 3.6b the open-circuit voltage (V_{oc}) is dominated by temperature.

3.3.2   Calculation the PV Boost DC/DC Converter

The output voltage from PV array has small value to be synchronized with the electrical grid through the DC/AC inverter. Therefore, the DC/DC boost converter is employed to step up the output voltage of PV array in order to achieve the required voltage level, as shown in Fig. 3.7. The configuration of the DC/DC boost converter

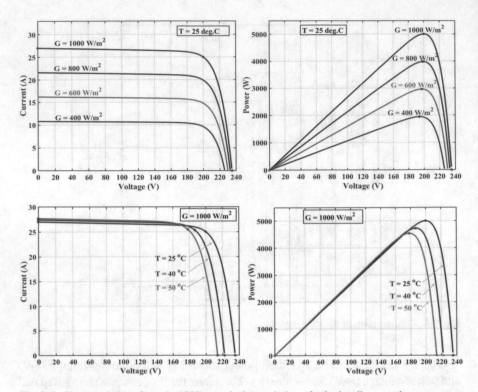

Fig. 3.6 Characteristics of a typical PV array during variation of solar irradiance and temperature. (**a**) Current-voltage (I-V) and power-voltage (P-V) characteristics of PV array under variable solar irradiance. (**b**) Current-voltage (I-V) and power-voltage (P-V) characteristics of PV array under variable temperature

includes two nearly ideal semiconductor switches such as diode and MOSFET and energy storage elements such as inductor and capacitor. The storage elements in the boost converter act as a low-pass filter to reduce the voltage ripple. An input capacitor (C_a) is employed to stabilize the terminal voltage of PV array caused by varying converter input current due to switching, while an output capacitor (C_1) acts as a low-pass filter to reduce the output voltage ripple [68].

The operation modes of the DC/DC boost converter are illustrated in Fig. 3.8. When the switch (Q) is turned on, current flows through the inductor (L_a) and switch (Q) in a clockwise direction, and the inductor stores some energy by generating a magnetic field ($V_{La} = V_{pv}$). When the switch (Q) is turned off, the magnetic field previously created will be obliterated to maintain the current flows toward the DC link, and also the polarity of the induced voltage across the inductor is reversed. Therefore, the inductor voltage (V_{La}) adds to the array voltage (V_{pv}),and the output voltage (V_{dc}) will be greater than the input voltage ($V_{dc} = V_{pv} + V_{La}$). Furthermore, the MPPT technique is implemented on the boost converter to capture the maximum power from PV array during variation of the solar irradiance. Therefore, the switching duty cycle of the boost converter (Dy) is generated by the MPPT technique.

As a PV cell is a current source, a capacitor (C_a) is estimated using Eq. (2.5) and interconnected in parallel to the output terminal of PV array, so that it can work as a voltage source to the DC/DC boost converter. The relations between input and output variables of the DC/DC boost converter and the values of its elements are expressed as follows [69]:

$$C_a = \frac{Dy * V_{PV}}{4 * \Delta V_{PV} * f_s^2 * I_{dc}}$$

(3.5)

$$Dy = 1 - \frac{V_{PV}}{V_{dc}}$$

(3.6)

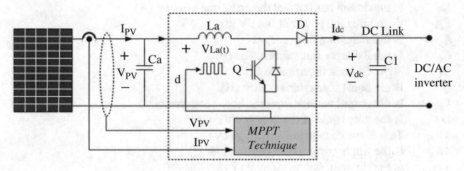

Fig. 3.7 Basic configuration of the DC/DC boost converter

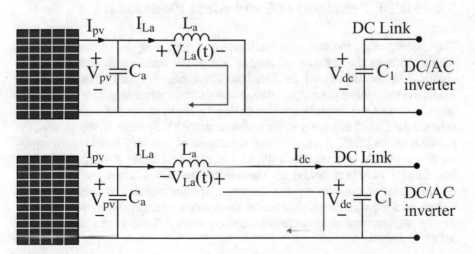

Fig. 3.8 Switching modes of the DC/DC boost converter. (**a**) Mode 1: when the switch (Q1) is turned on. (**b**) Mode 2: when the switch (Q1) is turned off

$$L_a = \frac{V_{PV} * (V_{dc} - V_{PV})}{\Delta I_{La} * f_S * V_{dc}} \qquad (3.7)$$

$$\Delta I_{La} = 0.13 * I_{PV} * \frac{V_{dc}}{V_{PV}} \qquad (3.8)$$

$$C_1 \geq \frac{P_{PV}}{\Delta V_0 * f_s * V_{dc}} \qquad (3.9)$$

where:

V_{PV} Is the input voltage to the converter from PV array (V)
I_{PV} Is maximum current that the array can provide (A)
P_{PV} Is nominal the power of the PV array (W)
f_s Is the switching frequency (Hz)
C_a Is the PV array link capacitance (F)
C_1 Is the DC link capacitance (F)
l_a Is the boost converter inductor (H)
V_{dc} Is the output voltage from the boost converter (V)
Dy   Is the duty cycle of the boost converter
ΔV_{PV} That is the change in PV voltage (V)
ΔI_{La} Is the ripple current of boost inductor (I)
ΔV_0 Is the ripple of output voltage (V)

3.4   MPPT Techniques of Stand-Alone PV System

The intensity of the incident solar irradiance on the PV array varies due to different reasons such as the variation of time in a day, the atmospheric effects such as clouds, and the latitude of the location. Therefore, the MPPT techniques are employed to regulate the output voltage and output current from the PV array in order to extract the maximum power during variation of the solar irradiance and enhance the overall efficiency of the grid-connected PV systems. In this section, the principle of the MPPT is a review, and simulation of two MPPT techniques implemented in the PV systems is introduced. Over the past decades, many methods to find the MPP have been developed. These techniques differ in many aspects such as required sensors, complexity, cost, the range of effectiveness, convergence speed, correct tracking when irradiation and/or temperature change, and hardware needed for the implementation or popularity, among others. Some of the most popular MPPT techniques are [14]:

1. Perturb and observe (hill climbing method)
2. Incremental conductance method

3. Fractional short circuit current
4. Fractional open circuit voltage
5. Fuzzy logic
6. Neural networks
7. Ripple correlation control
8. Current sweep
9. DC-link capacitor droop control
10. Load current or load voltage maximization
11. $\dfrac{dP}{dV}$ or $\dfrac{dP}{dI}$ feedback control

Among several techniques mentioned, the P&O method and the InCond algorithms are the most commonly applied algorithms. Other techniques based on different principles include fuzzy logic control, neural network, fractional open circuit voltage or short circuit current, current sweep, etc. Most of these methods yield a local maximum, and some, like the fractional open circuit voltage or short circuit current, give an approximated MPP, rather than an exact output. In normal conditions the V-P curve has only one maximum. However, if the PV array is partially shaded, there are multiple maxima in these curves. Both P&O and InCond algorithms are based on the "hill climbing" principle, which consists of moving the operation point of the PV array in the direction in which the power increases. Hill climbing techniques are the most popular MPPT methods due to their ease of implementation and good performance when the irradiation is constant. The advantages of both methods are simplicity and requirement of low computational power. The drawbacks are as follows: oscillations occur around the MPP, and they get lost and track the MPP in the wrong direction during rapidly changing atmospheric conditions. In the following, a review and simulation results of P&O and InCond MPPT techniques that are implemented in the PV systems are introduced.

3.4.1 Perturb and Observe MPPT Technique

The generated power and the output current from PV array vary nonlinearly with the array output voltage and the solar irradiance level. Therefore, it is essential to operate the PV array at the optimum voltage level (V_{MPP}) to extract the maximum power from it and increase the overall efficiency of the PV conversion systems. The maximum power point (MPP) is obtained when the gradient of power-voltage (P-V) curve is equal to zero as illustrated in Fig. 3.9. Thus, in order to track the MPP, the output voltage from PV array (V_{PV}) is regulated so that it increases when the derivative of power with respect to voltage is positive ($\dfrac{dP_{\text{pv}}}{dV_{\text{pv}}} > 0$), and it decreases when the derivative of power with respect to voltage is negative ($\dfrac{dP_{\text{pv}}}{dV_{\text{pv}}} < 0$). The control algorithm which provides continuous tracking of the MPP can be expressed as follows [70]:

$$V_{MPP} = K_1 \int \frac{dP_{pv}}{dV_{pv}}\, dt \qquad (3.10)$$

The algorithm involves a perturbation on the duty cycle of the power converter and a perturbation in the operating voltage of the DC link between the PV array and the power converter. Perturbing the duty cycle of the power converter implies modifying the voltage of the DC link between the PV array and the power converter. In this method, the sign of the last perturbation and the sign of the last increment in the power are used to decide the next perturbation. As can be seen in Fig. 3.9, on the left of the MPP incrementing the voltage increases the power, whereas on the right decrementing the voltage decreases the power. If there is an increment in the power, the perturbation should be kept in the same direction, and if the power decreases, then the next perturbation should be in the opposite direction. Based on these facts, the algorithm is implemented as shown in the flowchart in Fig. 3.10, and the process is repeated until the MPP is reached.

The P&O algorithm is one of the most popular MPPT techniques due to its simplicity, ease of implementation, and requirement of low computational power. The algorithm involves a perturbation on the duty cycle of the DC/DC converter that implies modifying the operating voltage of the DC link between the PV array and the DC/DC converter. In the P&O MPPT technique, the sign of last perturbation and the sign of the last increment in the power are used to decide the next perturbation. Therefore, if there is an increment in the power, the next perturbation should be kept in the same direction, and if the power decreases, then the subsequent perturbation should be in the opposite direction. Based on these facts, the P&O MPPT technique can be summarized in Table 3.1 [47].

The main drawback of the P&O MPPT technique is the oscillation around the MPP instead of directly tracking it. Since, when the operating point reaches very

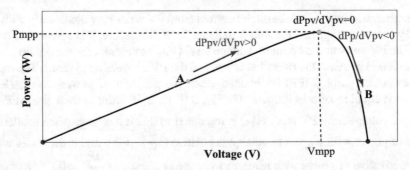

Fig. 3.9 The basic principle of MPPT in PV conversion systems

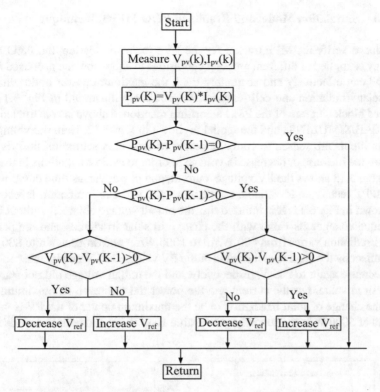

Fig. 3.10 Flow chart of the P&O MPPT technique

Table 3.1 Summary of the P&O MPPT technique

Perturbation	Change in power	Next perturbation
Positive	Positive	Positive
Positive	Negative	Negative
Negative	Positive	Negative
Negative	Negative	Positive

close to the MPP, it does not stop at the MPP and keeps on perturbing in both the directions. The oscillation can be minimized by reducing the perturbation step size. However, the smaller perturbation size slows down the response of the MPPT. The solution to this conflicting situation is to have a variable perturbation step size that gets smaller toward the MPP [71]. Moreover, the P&O MPPT technique cannot track the MPP during the lower solar irradiance levels and when the solar irradiance changes rapidly. During the rapid variation of the solar irradiance, this MPPT technique considers the change in the MPP is due to perturbation and ends up calculating the MPP in the wrong direction.

3.4.1.1   Simulation Model and Results of P&O MPPT Technique

In order to verify the MPP tracker for the PV system simulation, the P&O MPPT strategy is applied at different ambient conditions to show how the proposed MPPT method can effectively and accurately track the maximum power under different. The solar irradiation and cell temperature profile are illustrated in Fig. 3.11. The detailed block diagram of the P&O algorithm mentioned above is constructed using MATLAB/SIMULINK, and the model is shown in Fig. 3.12. Here the voltage and current inputs are sensed to compute power as shown. A saturation limit is set to monitor the increase or decrease in voltage in order to avoid oscillations in the MPP.

Figure 3.13 shows the PV voltage, current, and power versus time curve without the MPPT technique at variable temperature and variable irradiation levels which are shown in Fig. 3.11. It is inferred that the output voltage obtained without MPPT technique changes its value with the change in solar irradiation and temperature. Solar irradiation varies from 600 W/m² to 1000 W/m² and drops again to 800 W/m² to be offset by the change in the value of the PV voltage 195 V and increase to 225 V and decrease again to 205 V, respectively, and the output voltage did not start from the desired value. For the current and the power did not reach its maximum value with the change of solar irradiation to be the maximum power of the PV is 4000 W instead of 5000 W at the sun irradiation value 1000 W/m² and temperature of 25 °C.

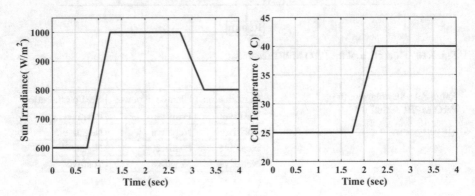

Fig. 3.11 The solar irradiation and cell temperature profile

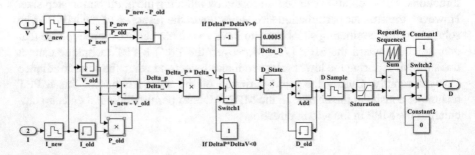

Fig. 3.12 MATLAB/Simulink model of the P&O MPPT technique

The P&O method was applied, and the output was observed as shown in Fig. 3.14 for the same profile of irradiation and temperature. As shown in Fig. 3.14, the output voltage of PV array traces the desired value well in response to the variation of the solar irradiance. When the solar irradiance is decreased from $G = 600$ W/m² to $G = 1000$ W/m², the MPPT controller increases the array output voltage from 200 V to 204 V and go back again to 200 V after irradiance constancy at 1000 W/m² in order to extract the maximum power from the PV array. Then, the MPPT controller decreases the output voltage of PV array from 200 V to 188 V, in response to the variation of temperature of PV array from 25 °C to 40 °C and change the solar irradiance from $G = 1000$ W/m² to $G = 800$ W/m². Therefore, the MPPT controller can accurately track the PV array voltage at the maximum power point (V_{mpp}) to harness the maximum power from the PV array during the rapid variation of solar irradiance and cell temperature. Also, Fig. 3.14 illustrates that the output current of PV array (I_{pv}) reflects the same scenario of the solar irradiance and cell temperature. When the solar irradiance is changed from $G = 600$ W/m² to $G = 1000$ W/m², it leads to increase the output current of PV array from 15 A to 25 A. Then, the PV array current decreases from 25 A to 20 A, in response to the change of temperature of PV

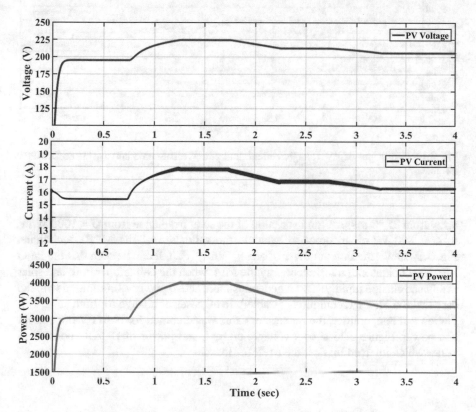

Fig. 3.13 The output of PV voltage, current, and power versus time curve without MPPT technique

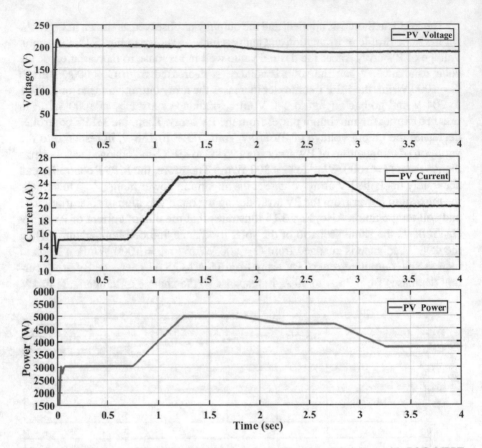

Fig. 3.14 The output of PV voltage, current, and power versus time curve with P&O MPPT technique

array from 25 °C to 40 °C and variation of the solar irradiance from $G = 1000$ W/m^2 to $G = 800$ W/m^2. In order to evaluate the validation of the MPPT technique, Fig. 3.14 shows the output power of one PV array (P_{pv}). It can be seen that the P&O MPPT technique can track accurately the MPP when the cell temperature and solar irradiance change rapidly; also it generates more active power as compared with the case that the MPPT technique is disabled. In all cases of change for both solar irradiation and temperature, the maximum value was obtained for both PV out current and power. Figure 3.15 shows voltage, current, and power of DC/DC boost converter which inferred that get the MPP for loads.

3.4.2   *Incremental Conductance MPPT Technique*

The InCond MPPT technique is widely implemented in the PV conversion systems due to its simplicity and advantage of offering good performance during the lower solar irradiance levels and when the solar irradiance changes rapidly. The InCond MPPT technique utilizes the current and voltage sensors to sense the output current and voltage of the PV array. In the InCond MPPT method, the array terminal

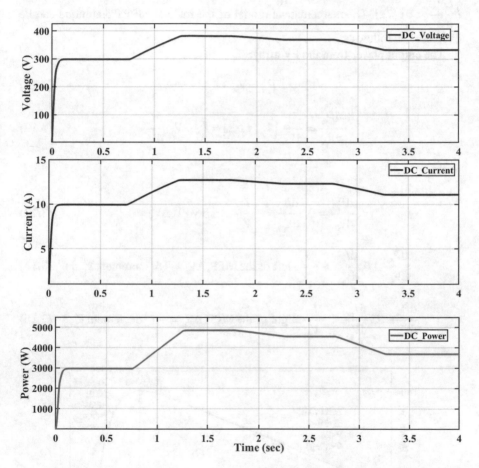

Fig. 3.15 The output DC/DC boost converter – voltage, current, and power with P&O MPPT technique

voltage (V_{PV}) is always adjusted according to the PV array voltage at MPP (V_{MPP}); it is based on the incremental and conductance of the PV array. The basic concept of the InCond MPPT technique is illustrated in Fig. 3.16.

The flow chart of this MPPT technique is shown in Fig. 3.17. The operation of InCond MPPT technique is based on the fact that the derivative of power with respect to voltage ($\frac{dP_{pv}}{dV_{pv}}$) is equal to zero at the MPP. Moreover, this derivative is positive at the left of the MPP ($\frac{dP_{pv}}{dV_{pv}} > 0$) and is negative at the right of the MPP ($\frac{dP_{pv}}{dV_{pv}} < 0$) [72]. The mathematical model of the InCond MPPT technique can be expressed as follows:

The output power from the PV array:

$$P_{pv} = V_{pv} * I_{pv} \tag{3.11}$$

$$\frac{dP_{pv}}{dV_{pv}} = \frac{d}{dV_{pv}}\left[V_{pv} * I_{pv}\right] = I_{pv} + V_{pv}\frac{dI_{pv}}{dV_{pv}} \tag{3.12}$$

Then,

$$\frac{dP_{pv}}{dV_{pv}} = 0, \frac{dI_{pv}}{dV_{pv}} = -\frac{I_{pv}}{V_{pv}} \text{ at the MPP}, \Delta V_n = 0 \tag{3.13}$$

$$\frac{dP_{pv}}{dV_{pv}} > 0, \frac{dI_{pv}}{dV_{pv}} > -\frac{I_{pv}}{V_{pv}} \text{ left of the MPP}, \Delta V_n = +\delta\left(\text{increment } V_{pv}\right) \tag{3.14}$$

$$\frac{dP_{pv}}{dV_{pv}} < 0, \frac{dI_{pv}}{dV_{pv}} < -\frac{I_{pv}}{V_{pv}} \text{ Right of the MPP}, \Delta V_n = -\delta\left(\text{decrement } V_{pv}\right) \tag{3.15}$$

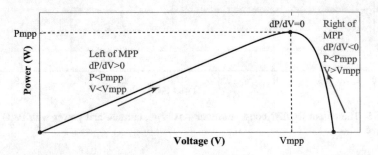

Fig. 3.16 The basic concept of InCond MPPT technique

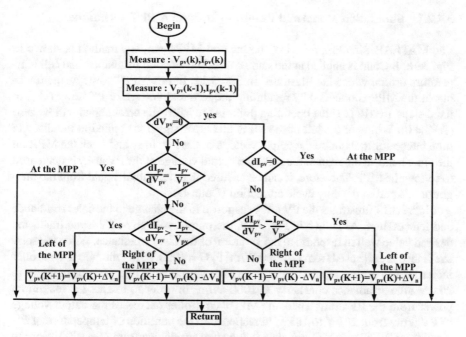

Fig. 3.17 Flow chart of InCond MPPT technique

Table 3.2 Major characteristics of the MPPT techniques

MPPT technique	PV array dependence	True MPP?	Convergence speed	Implementation complexity	Sensed parameters
P&O	No	Yes	Varies	Low	Voltage, current
InCond	No	Yes	Varies	Medium	Voltage, current

Thus, the MPP can be tracked by comparing the InCond $\left(\dfrac{dI_{pv}}{dV_{pv}}\right)$ with the instantaneous conductance $\left(\dfrac{I_{pv}}{V_{pv}}\right)$ as illustrated in the flowchart in Fig. 3.17. This algorithm increments or decrements the array terminal voltage (V_{PV}) to track the MPP during variation of the solar irradiance. The major characteristics of the presented MPPT techniques can be summarized in Table 3.2 [73].

3.4.2.1   Simulation Model and Results of InCond MPPT Technique

The MATLAB/Simulink model of the InCond MPPT strategy method is shown in Fig. 3.18. InCond is applied at the same conditions of solar irradiation and cell temperature profile which are illustrated in Fig. 3.11. In this MPPT strategy, the tracking of the MPP is obtained by regulating the terminal voltage of PV array (V_{PV}) in fixed steps $(\pm \Delta V_n)$. If the operating point is the MPP, the error signal will be zero ($\Delta V_n = 0$). While, at the left of the MPP, this error signal is applied to the discrete time integrator to increase the output voltage of the PV array and track the MPP, on the other hand, at the right of the MPP, the output voltage of the PV array is decreased to follow the MPP. Therefore, it can be ensured that the output signal from the integrator is equal to the duty cycle correction (Delta-D).

Figure 3.19 illustrates the PV array response to the change in the solar irradiance and temperature. As shown in Fig. 3.19, the output voltage of PV array traces the desired value well in response to the variation of the solar irradiance. When the solar irradiance is decreased from $G = 600$ W/m^2 to $G = 1000$ W/m^2, the MPPT controller increases the array output voltage from 201 V to 209 V and goes back again to 201 V after irradiance constancy at 1000 W/m^2 in order to extract the maximum power from the PV array. Then, the MPPT controller decreases the output voltage of PV array from 201 V to 185 V, in response to the variation of temperature of PV array from 25 °C to 40 °C and change the solar irradiance from $G = 1000$ W/m^2 to $G = 800$ W/m^2. Therefore, the MPPT controller can accurately track the PV array voltage at the maximum power point (V_{mpp}) to harness the maximum power from the PV array during the rapid variation of solar irradiance and cell temperature.

Also, Fig. 3.19 illustrates that the output current of PV array (I_{pv}) reflects the same scenario of the solar irradiance and cell temperature. When the solar irradiance is changed from $G = 600$ W/m^2 to $G = 1000$ W/m^2, it leads to increase the output current of PV array from 15 A to 25 A. Then, the PV array current decreases from 25 A to 20 A, in response to the change of temperature of PV array from 25 °C to 40 °C and variation of the solar irradiance from $G = 1000$ W/m^2 to $G = 800$ W/m^2. In order to evaluate the validation of the MPPT technique, Fig. 3.19 shows the output power of one PV array (P_{pv}). It can be seen that the InCond MPPT technique can track accurately the MPP when the cell temperature and solar irradiance change rapidly; also it generates more active power as compared with the case that the MPPT technique is disabled. Figure 3.20 shows voltage, current, and power of DC/DC boost converter which inferred that get the MPP for loads.

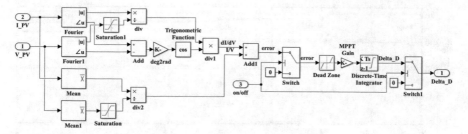

Fig. 3.18 MATLAB/Simulink model of the InCond MPPT technique

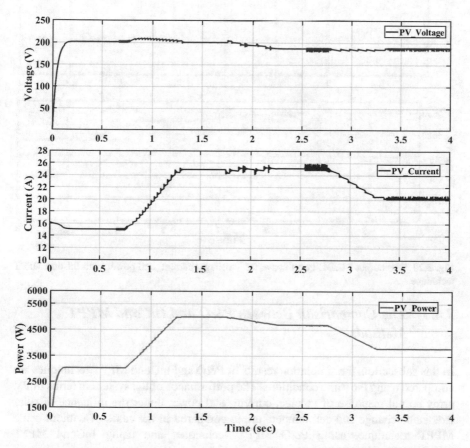

Fig. 3.19 Output of PV voltage, current, and power versus time curve with InCond MPPT technique

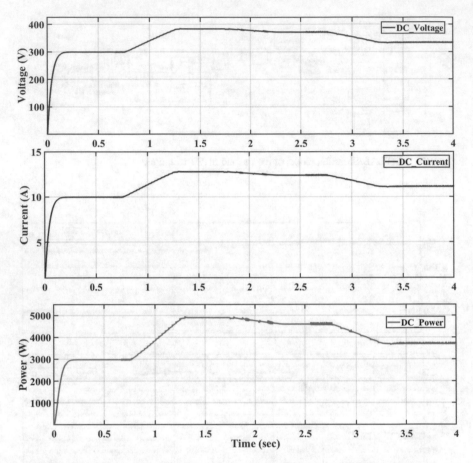

Fig. 3.20 The output DC/DC boost converter – voltage, current, and power with InCond MPPT technique

3.4.3   The Comparison Between P&O and InCond MPPT Methods

In this subsection, the simulation results of P&O and InCond MPPT techniques are compared using the same conditions. The performance of the system in terms of PV array output response of voltage, current, and power under the influence of solar irradiance change and cell temperature is compared in the cases of without using MPPT technique, using P&O MPPT technique, and using InCond MPPT technique.

Figure 3.21 illustrations of the output voltage of the PV array, with the beginning of the system work, notice that the voltage of P&O technique is reached to study state first at approximately 200 V and followed by InCond technique at the same voltage value. In the case of nonuse of the MPPT method, the voltage starts from a

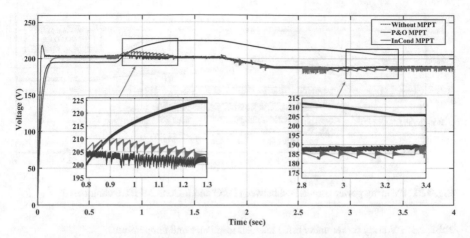

Fig. 3.21 PV array voltage comparison between P&O and InCond MPPT techniques

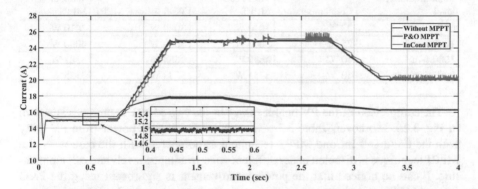

Fig. 3.22 PV array current comparison between P&O and InCond MPPT techniques

value less than the desired value (195 V). The difference between the results of the three cases is significant when a change in the value of solar irradiance occurs from $G = 600$ W/m^2 to $G = 1000$ W/m^2. We find that the PV output voltage of P&O technique is more stable, while the MPPT does not use the voltage to reach 225 V. The PV voltage produced by P&O and InCond MPPT technique is almost identical in the rest of the change to both solar irradiance and cell temperature.

The comparison of the PV output current in the three cases is shown in Fig. 3.22. It is clear from the figure that the current in P&O technique has more constancy and stability than the current in the InCond MPPT technique. The beginning of the system works of the PV current of P&O technique is reached to study state first at approximately 15A and followed by InCond technique at 15.1A. In the case of non-use of the MPPT method, the PV current starts from a value higher than the desired value (15.5 V). It also did not reach the maximum current when solar irradiance increased to 1000 W/m^2.

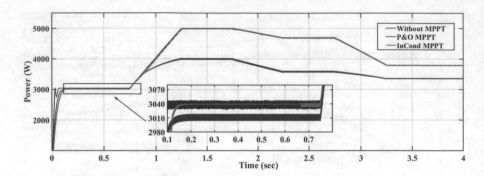

Fig. 3.23 PV array power comparison between P&O and InCond MPPT techniques

Table 3.3 PV array power under different solar irradiance and temperature

Solar irradiance level	Cell temperature	PV array power		
		With P&O MPPT	With InCond MPPT	Without MPPT
600 W/m²	25 °C	3040 W	3040 W	3010 W
1000 W/m²	25 °C	5000 W	5000 W	4000 W
1000 W/m²	40 °C	4690 W	4685 kW	3585 W
800 W/m²	40 °C	3785 W	3788 W	3355 W

The comparison of the PV output power in the two MPPT techniques is shown in Fig. 3.23. Moreover, Table 3.3 demonstrates the supply power from PV array with the P&O and InCond MPPT technique as compared with the case when the MPPT technique is disabled under different solar irradiance levels and cell temperature. It can be noticed that the power improvement is significant using the P&O InCond MPPT technique as compared with the case when the MPPT technique is disabled (fixed duty cycle of 35%).

3.5　Summary

This chapter presented the operating principle of the PV conversion systems that generate electricity via the PV effect, in which semiconductor holes and electrons are freed by photons from the incident solar irradiance. The PV systems are equipped with the DC/DC converter to implement the MPPT technique. Furthermore, this chapter introduced a review of two MPPT techniques that are implemented in the PV systems. The P&O MPPT technique and InCond MPPT technique are the most commonly implemented PV conversion systems due to its simplicity and advantage of offering good performance when the solar irradiance changes rapidly. The two MPPT techniques were simulated by the MATLAB/Simulink, and the results response of the PV array from voltage, current, and power are compared to the effect of solar irradiation and temperature change.

Chapter 4
Improving the Resiliency of a PV Stand-Alone with Energy Storage

4.1 Introduction

The stand-alone PV system is controlled using MPPT algorithm under the impact of the fixed or changing solar irradiation, and the system is connected to BES to produce power for variable AC loads. The system comprises of PV array, BES, DC/DC boost converter circuit, single-phase inverter with LCL filter, and bidirectional DC/DC buck-boost converter performing as charging circuit. The MPP be able to obtain by controlling the duty cycle fed to the gate of the IGBT transistor located within the boost circuit. Moreover, the control method used with both the single-phase inverter and the buck-boost converter circuit is a dynamic error driven PI controller.

This chapter investigates a dynamic modeling, simulation, and control strategy of the proposed PV stand-alone system with BES. Moreover, this chapter discusses the performance comparison of PV stand-alone system with BES in two cases of operation. In the first case, the system operates without and with BES under constant solar irradiation. In the second case, the PV system is connected to a BES and operates under a variable in solar irradiation. In addition to presenting the operating results of the system for the two cases mentioned previously, harmonic analysis is performed on these results using the fast Fourier transform (FFT) tools.

4.2 Structure and Modeling a PV Stand-Alone with Battery Energy Storage

The objective of this section is to design an independent PV system containing a PV panel, DC/DC boost converter, and half-bridge buck-boost bidirectional chopper and focuses on increasing energy extraction by improving MPPT. The P&O MPPT technique is implemented on the DC/DC boost converter to extract the maximum

© Springer Nature Switzerland AG 2019
A. A. Elbaset et al., *Performance Analysis of Photovoltaic Systems with Energy Storage Systems*, https://doi.org/10.1007/978-3-030-20896-7_4

power from the PV station during variation of the solar irradiance. The system is rated to 5 kW generated by PV array and Two loads; each of 3 kW can be connected to the DC/AC inverter. This makes the overall electrical power consumed 6 kW when both are connected. Figure 4.1 clarifies the simplified diagram of the stand-alone PV system with BES. The mathematical model for each of the system components is described in the following subsections.

4.2.1   Mathematical Modeling of the PV Array Under Study

The studied PV array is formed by connecting a number of 4 PV strings in parallel in order to increase the output current and achieve the array rated power ($P_{array} = 5040$ W). The PV string is composed by connecting number of 6 PV modules in series to increase the output voltage at the MPP ($V_{mpp} = 202.8$ V). Figure 4.2 shows the electrical modeling of the PV array based on the Shockley diode. Also, the implementation of the PV array in MATLAB/Simulink model is depicted in

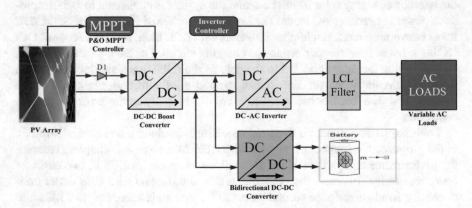

Fig. 4.1 Simplified diagram of the stand-alone photoelectric system with energy storage

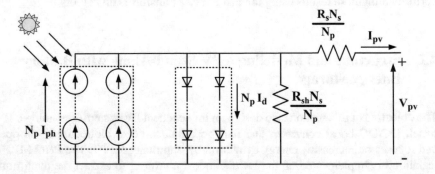

Fig. 4.2 The electrical model of the PV array

Fig. 4.3. The electrical characteristics of the PV array can be simulated with regard to the variations in the environmental conditions such as the solar irradiance intensity and temperature.

The current-voltage (I-V) characteristics and the power-voltage (P-V) characteristics of the employed PV array during variation of the solar irradiance and the corresponding current-voltage (I-V) relationship of the PV array have been introduced and reviewed in Sect. 2.3.1. Furthermore, the detailed specifications of the studied PV array are listed in Table 4.1.

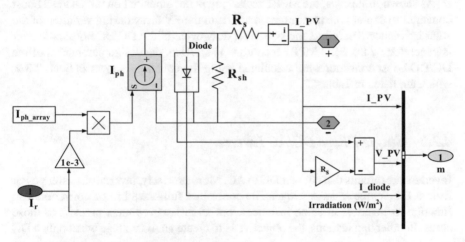

Fig. 4.3 MATLAB/Simulink model of the PV array

Table 4.1 Design parameters of the PV array

Design parameter	Symbol	Value
Module type	Solaria Solaria 210	
Rated power of PV module	$P_{PV-module}$	209.96 W
Number of cells per module	N_{ser}	70
Number of parallel strings	N_P	4
Number of series-connected modules	N_S	6
Light-generated current	I_{pv}	7.1774 A
PV saturation current	I_s	0.27907 nA
Series resistance	R_s	0.40587 Ω
Parallel resistance	R_{sh}	55.9684 Ω
Maximum power per array	P_{array}	5040 W
Voltage at maximum power point of PV array	V_{mpp}	202.8 V
Current at maximum power point of PV array	I_{mpp}	24.852 A
Open-circuit voltage of PV module	V_{oc}	41.59 V
Short-circuit current of PV module	I_{sc}	7.13 A

4.2.2 DC/DC Boost Converter

Two-stage topology selected for the DC-DC chopper in PV system takes care of MPPT. The employment of two-stage topology allows the system to be customizable, i.e., it can be transformed to a multi-string system to increase the capacity of the system in the future, with each series having its own MPPT and DC/DC boost converter. This was about the significance of the DC/DC converter, but the configuration is illustrated in Fig. 4.4.

As shown in Fig. 4.4, the MPPT technique is implemented on the DC/DC boost converter to capture the maximum power from the PV array during variation of the solar irradiance. Therefore, the switching duty cycle of the DC/DC boost converter is generated by the P&O MPPT technique [71]. Also, the design parameters of the DC/DC boost converter were calculated using the equations shown in Sect. 2.3.2, which are listed in Table 4.2.

4.2.3 Single-Phase DC/AC Inverter

Inverters are circuits that convert DC to AC. More precisely, inverters transfer power from a DC source to an AC load. The controlled full-wave bridge converters can function as inverters in some instances, but an AC source must preexist in those cases. In other applications, the objective is to create an AC voltage when only a DC

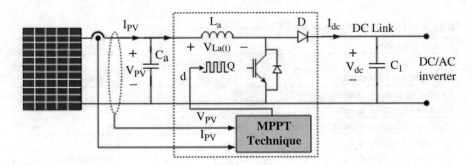

Fig. 4.4 The PV array with the DC/DC boost converter

Table 4.2 Design parameters of the DC/DC boost converter

Design parameter	Symbol	Value
Input capacitance	C_a	1 mF
Inductance of boost converter	L_a	240 µH
Output capacitance	C_1	5 mF
Switching frequency	f_s	10 kHz
Output voltage	V_{dc}	280 V

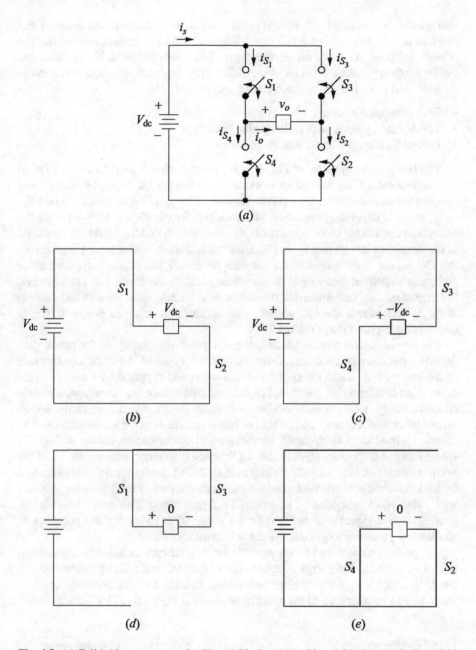

Fig. 4.5 (a) Full-bridge converter; (b) S1 and S2 closed; (c) S3 and S4 closed; (d) S1 and S3 closed; (c) S2 and S4 closed

voltage source is available. The focus of this section is on single-phase inverters that produce an AC output from a DC input. Inverters are used in applications such as adjustable-speed AC motor drives, uninterruptible power supplies, PV systems, and running AC appliances from an automobile BES. Inverters are power electronic devices used in various PV system configurations [24, 74]:

- Grid-connected systems
- Stand-alone systems with rechargeable batteries
- Pumping systems without storage batteries

The full-bridge converter of Fig. 4.5a is the basic circuit used to convert DC to AC. In this circuit, an AC output is synthesized from a DC input by closing and opening the switches in an appropriate sequence. The output voltage V_O can be $+V_{dc}$, $-V_{dc}$, or zero, depending on which switches are closed. Figure 4.5b–e shows the equivalent circuits for switch combinations. Note that S_1 and S_4 should not be closed at the same time nor should S_2 and S_3. Otherwise, a short circuit would exist across the DC source. Real switches do not turn on or off instantaneously. Therefore, switching transition times must be accommodated in the control of the switches. Overlap of switch "on" times will result in a short circuit, sometimes called a shoot-through fault, across the DC voltage source. The time allowed for switching is called blanking time [69] (Table 4.3).

The power quality required for loads must match the quality of the power produced by the inverter. Therefore, there are different types of inverters. The function of the inverter is to attach the systems to each other and supply the PV energy in the network as efficiently as possible. For the required square wave voltage, a simple control strategy is employed to obtain the inverter switch signals, which are turned on and off at AC frequency and consist of high consistent currents and voltages. The control method used is dynamic error-driven PI controller as shown in Fig. 4.6 where its input is the error signal resulting from the difference between the V_{dc} of the boost converter and the reference voltage ($V_{ref} = 280V$). The control signal generated by the PI controller is adjusted to be a value in the range (-1 to 1) in order to match with a sine wave generated by sine wave (V_{ref}) generation. This gives a 60 Hz frequency control signal then utilized as the modulation index for the pulse width modulation generator to get the switching control signals.

The power that is used by converters is very significant in increasing the transport of power from PV energy system to the grid or AC loads. The harmonic results due to the operation of power electronic converters. The harmonic voltage and current ought to be restricted to the acceptable level for the point of PV generator link

Table 4.3 Switches cases of full-bridge inverter

Switches closed	Output voltage V_O
S1 and S2	$+V_O$
S3 and S4	$-V_O$
S1 and S3	0
S2 and S4	0

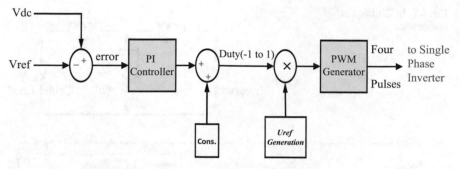

Fig. 4.6 The dynamic error-driven PI controller for the DC-AC inverter

to the networks and electrical loads. To ensure the harmonic voltage within the limit, each method of obtaining harmonic current can allow only a limited contribution, as per the international standard guideline. The filter must be used to reduce the total harmonics in AC voltage and current.

4.2.4 Filter Design

Recently, the development of renewable energy technologies has been accelerating, making the simultaneous development of power conversion devices for applications, such as wind and PV systems, extremely important; the development of these technologies is actively underway. The harmonics caused by the switching of the power conversion devices are the main factor-causing problems to sensitive equipment or the connected loads, especially for applications above several kilowatts, where the price of filters and total harmonics distortion (THD) is also an important consideration in the systems design phase [75]. The inductance of the input or output circuits of the power conversion devices has conventionally been used to reduce these harmonics. However, as the capacity of the systems has been increasing, high values of inductances are needed, so that realizing practical filters has been becoming even more difficult due to the price rises and the poor dynamic responses.

An L filter or LCL filter is usually placed between the inverter and the grid to attenuate the switching frequency harmonics produced by the grid-connected inverter. Compared with L filter, LCL filter has better attenuation capacity of high-order harmonics and better dynamic characteristic. However, an LCL filter can cause stability problems due to the undesired resonance caused by zero impedance at certain frequencies. To avoid this resonance from contaminating the system, several damping techniques have been proposed. One way is to incorporate a physical passive element, such as a resistor in series with the filter capacitor [76]. This passive damping solution is very simple and highly reliable. However, the additional resistor results in power loss and weakens the attenuation ability of the LCL filter. This drawback can be overcome by employing active damping (Figs. 4.7 and 4.8).

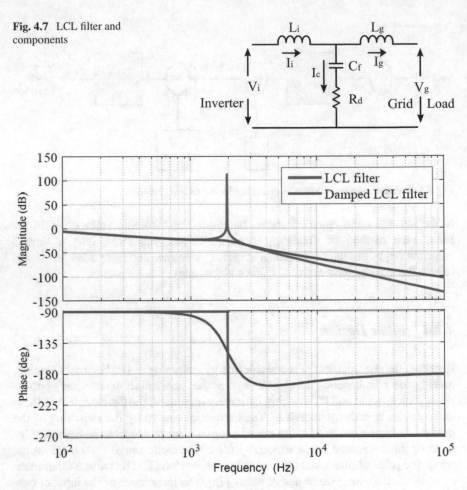

Fig. 4.7 LCL filter and components

Fig. 4.8 LCL filter with passive damping resistance [79]

The cutoff frequency (f_{res}) must have a sufficient distance from the grid frequency or the connected electrical AC loads. The cutoff frequency (f_{res}) of the LCL filter can be calculated as [77]

$$f_{res} = \frac{1}{2\pi} \times \sqrt{\frac{L_i + L_g}{L_i \times L_g \times C_f}} \qquad (4.1)$$

The LCL filter will be vulnerable to oscillations and it will magnify frequencies around its cutoff frequency (f_{res}). Therefore, the filter is added with damping. The simplest way is to add damping resistor (R_d). The variant with resistor connected in series with the filter capacitor has been chosen. The passive damped LCL filter frequency response is shown in Fig. 4.9. However, it is obvious that the damping

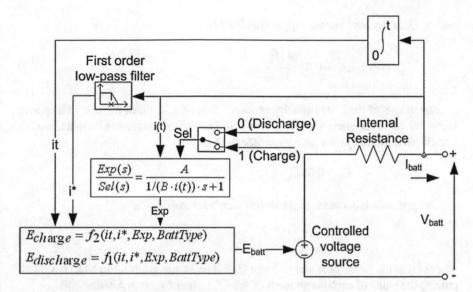

Fig. 4.9 The equivalent circuit of battery [81]

resistor reduces the efficiency of the overall system. The value of the damping resistor (R_d) can be calculated as [78]

$$R_d = \frac{1}{3\omega_{res}C_f}$$ (4.2)

The following parameters are needed for the filter design:

U_n, line-to-line RMS voltage (inverter output); P_n, rated active power; V_{dc}, DC link voltage; f_n, grid frequency; f_s, switching frequency. Thus, the filter values will be referred to in a percentage of the base values [79]:

$$Z_b = \frac{U_n^2}{P_n}$$ (4.3)

$$C_b = \frac{1}{\omega_n Z_b}$$ (4.4)

The first step in calculating the filter components is the design of the inverter side inductance (L_i), which can limit the output current ripple by up to 10% of the nominal amplitude. It can be calculated according to the equation derived in

$$L_i = \frac{V_{dc}}{16 f_s \Delta I_L}$$ (4.5)

where ΔI_L is the 10% current ripple specified by

$$\Delta I_L = 0.01 \frac{P_n \sqrt{2}}{U_n} \tag{4.6}$$

The design of the filter capacity proceeds from the fact that the maximal power factor variation acceptable by the grid is 5%. The filter capacity can be calculated as a multiplication of system base capacitance (C_b):

$$C_f = 0.05 C_b \tag{4.7}$$

The grid side inductance (L_g) can be calculated as

$$L_g = r * L_i \tag{4.8}$$

where (r) is the factor between (L_i) and (L_g). The M-file program in MATLAB calculates the value of each component of the LCL filter found in Appendix B.

4.2.5   Modeling of Battery Energy Storage

The BES is used to store solar energy whenever it is available in excess. Stored energy is used when PV energy is not enough to afford load demand. In addition, a complete review of the types of batteries used with PV systems was conducted in the third chapter. The focus was also on the type of lead-acid batteries. In this book, the BES is used from lead-acid type because it is more proper for renewable systems because of its limited cost and availability in large volume. Today lead-acid batteries are the best effectual solution for independent renewable energy systems because of their low cost, deep cycling, high discharges, and recycling. The BES block implements a generic dynamic model parameterized to represent most popular types of rechargeable batteries. The Fig. 4.9 shows the BES equivalent circuit that the block models [80].

In this section, lead-acid BES model is implemented using Simulink. The corresponding equations for charge and discharge model are represented according to the following equations [14]:

Charge model ($i* < 0$)

$$f_2\left(\mathrm{it}, i^*, i, \mathrm{Exp}\right) = E_0 - K \cdot \frac{Q_b}{\mathrm{it} + 0.1 \cdot Q_b} \cdot i^* - K \cdot \frac{Q_b}{Q_b - \mathrm{it}} \cdot \mathrm{it} + \mathrm{Laplace}^{-1}\left(\frac{\mathrm{Exp}(s)}{\mathrm{Sel}(s)} \cdot \frac{1}{s}\right) \tag{4.9}$$

Discharge model ($i* > 0$)

$$f_1\left(\text{it},i^*,i,\text{Exp}\right) = E_0 - K \cdot \frac{Q_{\text{b}}}{\text{it}+0.1\cdot Q_{\text{b}}} \cdot i^* - K \cdot \frac{Q_{\text{b}}}{Q_{\text{b}}-\text{it}} \cdot \text{it} + \text{Laplace}^{-1}\left(\frac{\text{Exp}(s)}{\text{Sel}(s)} \cdot 0\right) \quad (4.10)$$

where

i	The battery current (A)
$i*$	The low-frequency current dynamics (A)
it	The extracted capacity (Ah)
Exp(s)	Exponential zone dynamics (V)
E_0	The constant voltage (V)
K	The polarization constant (Ah^{-1}) or polarization resistance (Ω)
Q_{b}	Maximum battery capacity (Ah)
Sel(s)	Represents the battery mode. Sel(s) = 0 during battery discharge, Sel(s) = 1 during battery charging

The BES discharge characteristics typical curve is consisting of three sections as shown in Fig. 4.10. The first section represents the exponential voltage drop when the BES is charged. The width of the drop depends on the BES type. The second section represents the charge that can be extracted from the BES until the voltage drops below the battery nominal voltage. The third section represents the total discharge of the BES, when the voltage drops rapidly.

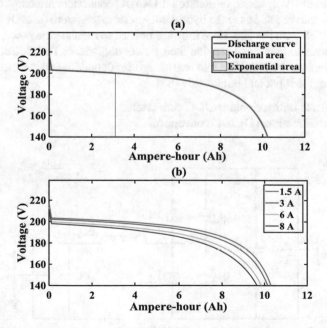

Fig. 4.10 Discharge characteristics of a lead-acid battery. (**a**) Nominal current discharge characteristic at (2A). (**b**) Discharge characteristic at diverse current values

4.2.6 Half-Bridge Bidirectional DC/DC Buck/Boost Converter

Bidirectional DC/DC converters serves the purpose of stepping up or stepping down the voltage level between its input and output along with the capability of power flow in both directions. Bidirectional DC/DC converters have attracted a great deal of applications in the area of the energy storage systems for hybrid vehicles, renewable energy storage systems, uninterruptable power supplies, and fuel cell storage systems. Bidirectional DC/DC converters are employed when the DC bus voltage regulation must be achieved along with the power flow capability in both the directions. One such example is the power generation by wind or PV power systems, where there is a large fluctuation in the generated power because of the large variation and uncertainty of the energy supply to the conversion unit (PV array) by the primary source [82]. These systems cannot serve as a stand-alone system for power supply because of these large fluctuations and therefore these systems are always backed up and supported by the auxiliary sources which are rechargeable such as BES units or SCs. These sources supplement the main system at the time of energy deficit to provide the power at a regulated level and get recharged through the main system at the time of surplus power generation or at their lower threshold level of discharge. Bidirectional DC/DC converter is needed to be able to allow power flow in both the directions at the regulated level.

Likewise, in PV systems, bidirectional DC/DC converters are employed to link up the high-voltage DC bus to the hybrid energy storage system (usually a combination of the BES with the *SC*). Here they are needed to regulate the power supply for electrical loads to help them provide their power demanded. The bidirectional DC/DC converters can be classified into two categories depending on the galvanic isolation between the input and output side [83]:

- Non-isolated bidirectional DC/DC converters.
- Isolated bidirectional DC/DC converters.

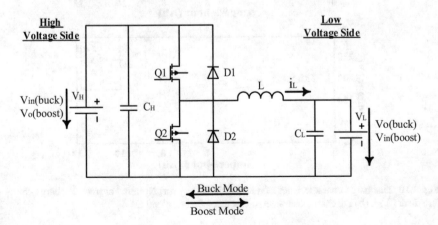

Fig. 4.11 Circuit diagram of half-bridge bidirectional DC/DC converter

In this book, the non-isolated bidirectional half-bridge DC/DC converters are used to connect the BES with the PV system, which illustrated its circuit diagram in Fig. 4.11. Basically, a non-isolated bidirectional DC/DC converter can be derived from the unidirectional DC/DC converters by enhancing the unidirectional conduction capability of the conventional converters by the bidirectional conducting switches. Due to the presence of the diode in the basic buck and boost converter circuits as shown in Fig. 4.12a, b, they do not have the inherent property of the bidirectional power flow. This limitation in the conventional boost and buck converter circuits can be removed by introducing a power MOSFET or an IGBT having an antiparallel diode across them to form a bidirectional switch and hence allowing current conduction in both directions for bidirectional power flow in accordance with the controlled switching operation as clarified in Fig. 4.11. When the Buck and the boost converters are connected in antiparallel across each other with the resulting circuit is basically having the same structure as the fundamental Boost and Buck structure but with the added feature of bidirectional power flow [58, 60].

The above circuit can be made to work in buck or boost mode depending on the switching of the switches Q1 and Q2. The switches Q1 and Q2 in combination with the antiparallel diodes D1 and D2 (acting as a freewheeling diode), respectively, make the circuit step up or step down the voltage applied across them. The bidirectional operation of the above circuit can be explained in the below two modes as follows:

Mode 1 (Boost Mode) In this mode, switch Q2 and diode D1 enter into conduction depending on the duty cycle, whereas the switch S1 and diode D2 are off all the time. This mode can further be divided into two intervals depending on the conduction on the switch 1 and diode D2 as shown in Fig. 4.12b.

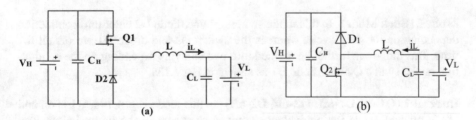

Fig. 4.12 (a) Buck converter circuit and (b) boost converter circuit

Interval 1 (Q2-on, D2-off; Q1-off, D2-off) In this mode Fig. 4.13a, Q2 is on and hence can be considered to be short-circuited; therefore, the lower-voltage battery charges the inductor, and the inductor current goes on increasing till not the gate pulse is removed from the Q2. Also, since the diode D1 is reversed biased in this mode and the switch S1 is off, no current flows through the switch Q1.

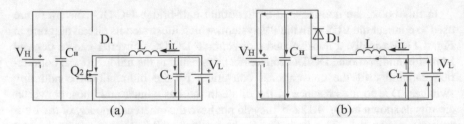

Fig. 4.13 Boost mode. (**a**) Interval 1, (**b**) Interval 2

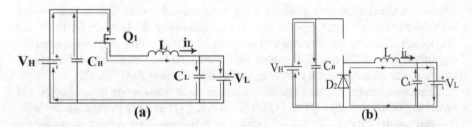

Fig. 4.14 Buck mode. (**a**) Interval 1, (**b**) Interval 2

Interval 2 (Q1-off, D1-on; Q2-off, D2-off) In this mode Fig. 4.13b, Q2 and Q1 both are off and hence can be considered to be open-circuited. Now since the current flowing through the inductor cannot change instantaneously, the polarity of the voltage across it reverses, and hence it starts acting in series with the input voltage. Therefore, the diode D1 is forward biased, and hence the inductor current charges the output capacitor C1 to a higher voltage. Therefore, the output voltage boosts up.

Mode 2 (Buck Mode) In this mode switch Q1 and diode D2 enter into conduction depending on the duty cycle, whereas the switch Q2 and diode D1 are off all the time. This mode can further be divided into two intervals depending on the conduction on the switch Q2 and diode D1 as shown in Fig 4.13a.

Interval 1 (Q1-on, D1-off; Q2-off, D2-off) In this mode Fig. 4.14a, Q1 is on and Q2 is off and hence the equivalent circuit is as shown in the figure below. The higher-voltage battery will charge the inductor and the output capacitor will get charged by it.

Interval 2 (Q1-off, D1-off; Q2-off, D2-on) In this mode Fig. 4.14b, Q2 and Q1 both are off. Again, since the inductor current cannot change instantaneously, it gets discharged through the freewheeling diode D2. The voltage across the load is stepped down as compared to the input voltage.

4.3 Simulation Results and Discussion of Stand-Alone PV System with BES

In this section, the dynamic performance of the PV system with BES during constant and variation of the solar irradiance is investigated. The temperature of PV array surface is considered to be constant at 25 °C during the entire simulation period. The PV array is rated 5 kW; MPPT control takes maximum power from PV using unidirectional DC/DC converter which is performed. The model designed in the previous parts is simulated in a Simulink environment which is illustrated in Fig. 4.15. The constructions of the system model to study the two cases are under constant and variable solar irradiation in different operating conditions. The chapter concentrates on the study of the impact of constancy or change in solar irradiation on the performance of BES. In the first case, the offered model works with constant solar irradiation and compares the results to the system in a case with and without BES. In the second case, the system works with variable solar irradiation. In both cases, the system is running in the same sequence. Initially, Load1 (3 kW) is run (*on*) and consumes more than half of the power generated by the PV system. After 1 second Load2 is running (*on*) making the total power consumed 6 kW. The PV system does not provide sufficient power; thus, supplemental energy is supplied by BES. This section also presents a consistent analysis of voltage and current after the filter using FFT tool. The load profile is connected to the system which is applied in all operating cases of the system as shown in Fig. 4.16. In the first operation case, a constant solar irradiation value (1000 W/m²) is applied to the system. In the second operation case, the system is employed into the variable value of solar irradiation which appears in Fig. 4.21. The simulation is run for 4 sec. In the next section, the operating results of the model on the MATLAB/Simulink program for both cases and the effect on current, voltage, power, and harmonic analysis by FFT tools are presented.

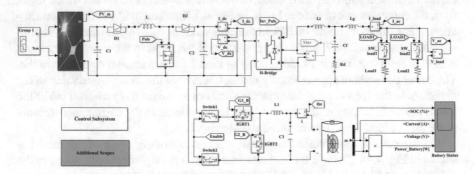

Fig. 4.15 Simulink model of a stand-alone PV system with a BES

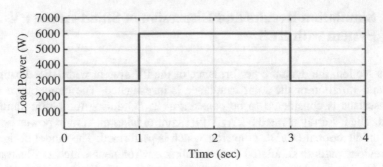

Fig. 4.16 Load profile

4.3.1 *Comparison Between PV System with and Without BES Under Constant Irradiation*

This subsection tackles the comparison of the operating results of the model in both cases with and without BES connected to the model. Comparison of the results of both cases is the focus of a set of important points: the amount of consumption when the load 1 is only in *on* situation, the entry and exit of the load 2, and the status of the system when there are loads that consume power more than what is generated by PV.

The current response of the PV system to the change in the load profile and with disconnecting and connecting the BES is illustrated in Fig. 4.17. As shown in Fig. 4.17a, there is no significant distinction in the current values of the PV for both cases. However, the output current of PV array (I_{pv}) is fairly constant at a value of 25 A if the BES is connected. At the start of operation, we notice that the current took a long time to settle at 25 A if no BES is connected to the system. However, it is observed that the current level is more stable when BES is used. Moreover, the existence of BES overcomes the effect of overshoot transient moments which result from the entry and exit of sudden loads. In Fig. 4.17b, when load 1 is only in *on* situation, it consumes a current of 19A in the case without BES. While the BES is connected, it consumes 14.7 A. This gives a 29.25% progress in the current. Transient moments resulting from the sudden entry of load 2 cause a current overshoot in the. The current overshoot approximately reaches 30.5% of the steady-state value in the first case. In the second case, the current overshoot approximately reaches 1% of the steady state of stability of the current. This improvement in the current also appears at the exit of load 2 and overcomes the undershoot.

Comparison of the voltage results for the two operating cases of the model is shown in Fig. 4.18. The result concludes that there is no significant difference in the PV voltage as illustrated in Fig. 4.18a but the overshoot at transient moments of the entry, exit of the load 2, and the starting of the model have been reduced. Illustrations, (b) and (c) of Fig. 4.18, show DC voltage and AC RMS load voltage which DC voltage (V_{dc}) stable at 280 V in the use of BES by suppressing or smoothing out transients that occur in PV systems.

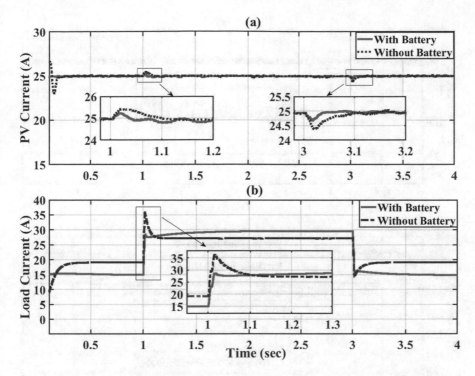

Fig. 4.17 The current response of PV system. (**a**) PV output current and (**b**) AC RMS load current

In the second case, when the load profile is changed from 3000 W to 6000 W, it leads to decrease in the DC voltage from 360 V to 256 V. Then, the DC voltage increases from 256 V to 360 V, in response to the change of the load profile from 6000 W to 3000 W, as illustrated in Fig. 4.18b. The AC voltage response at running the PV system without connecting BES are presented in Fig. 4.18c. The AC load voltage is decrease from 254.5 V to 181 V, when the load profile is changed from 3000 W to 6000 W. Then, the AC load voltage increases from 181 V to 254.5 V, in response to the change of the load profile from 6000 W to 3000 W.

The differences in current and voltage in two operating cases also appear when comparing the results of power as shown in Fig. 4.19. In order to evaluate the validation of the MPPT technique, Fig. 4.19a shows the output power of PV array (P_{pv}). It can be seen that the P&O MPPT technique can track accurately the MPP at 1000 W/m^2 of solar irradiance. Moreover, the impact of voltage and voltage stability is noticed during the use of the BES in power and provides the power consumed by each load according to need as much as possible. Relatively, the amount of power consumed in the system approaches 6 KW which is greater than what is generated by the PV array as shown in Fig. 4.19b.

The BES response for current, voltage, and SOC under the effect of the change in the load profile is shown in Fig. 4.20. When the model starts, the initial value of

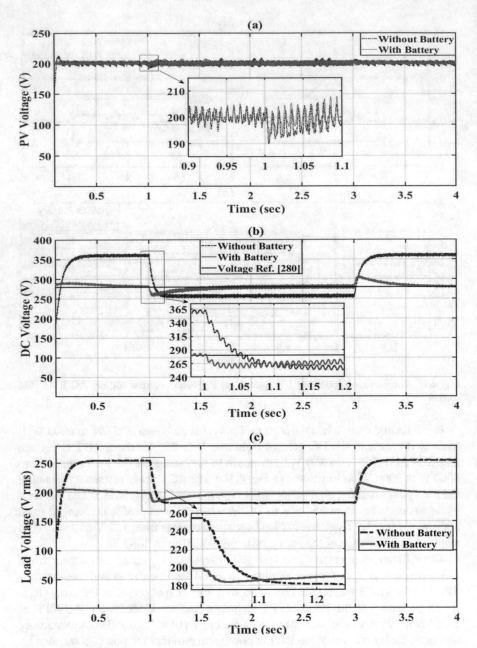

Fig. 4.18 The voltage response of PV system. (**a**) PV output voltage, (**b**) DC voltage, and (**c**) AC RMS load voltage

the profile is 3000 W, which is more than half of the power produced by the PV array, and thus the remaining power is about 2000 W. The remaining energy is charged by the BES as shown in Fig. 4.20a. When the load profile is changed from 3000 W to 6000 W, and this power is greater than what the PV array produce, the BES status switches from charge to discharge to compensate for the difference between the power required for the load and the power generated from the PV array. The change in BES status from charge to discharge shows an effect on SOC and voltage as illustrated in Fig. 4.20a, c. The change in the BES status effects on the battery current as its polarity changes from negative (charge) to positive (discharge) as shown in Fig. 4.20b. Then, the SOC and voltage increase again in response to the

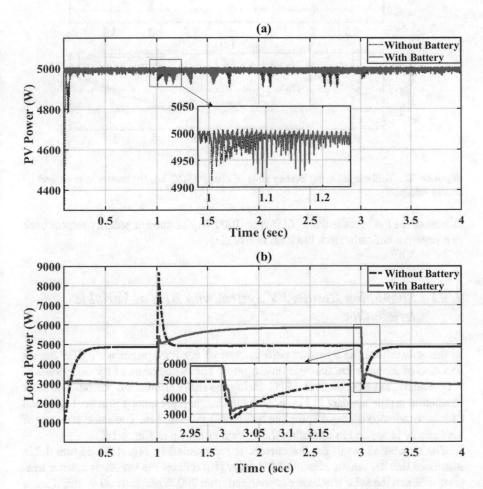

Fig. 4.19 The power response of PV system, (**a**) PV generated power, and (**b**) load active power

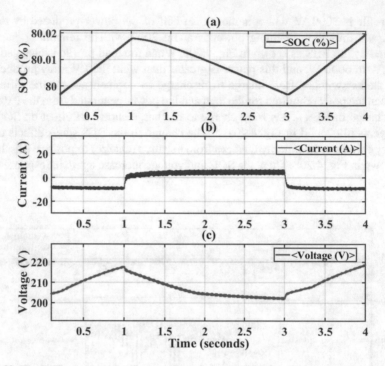

Fig. 4.20 The BES response. (**a**) Battery state of charge (SOC %), (**b**) battery current, and (**c**) battery voltage

change of the load profile from 6000 W to 3000 W. The current polarity returns back to a negative indication that the BES is charging.

4.3.2 Simulation Results PV System with BES at Variable Irradiation

In this subsection, the dynamic performance of the PV system with BES during variation of the solar irradiance is investigated. The temperature of PV array surface is considered to be constant at 25 °C during the entire simulation period. The solar irradiation profile appears in Fig. 4.21. This change represents a practical variation of solar irradiance during a complete one day as proposed in. The same scenario of load profile is applied in operating the system as shown in Fig. 4.16.

The simulation result for the current is represented in Fig. 4.22. Figure 4.22a illustrates that the output current of PV array (I_{pv}) reflects the variation in solar irradiance. When the solar irradiance is changed from 700 W/m^2 to 1000 W/m^2, it leads to increase in the output current of PV array from 17.5 A to 25 A. Then, the PV array current decreases from 25 A to 17.5 A, in response to the change of the solar irradiance from 1000 W/m^2 to 700 W/m^2. Variations in the AC curve Fig. 4.22b due to the

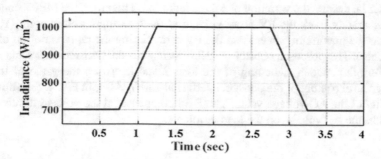

Fig. 4.21 Solar irradiation profile

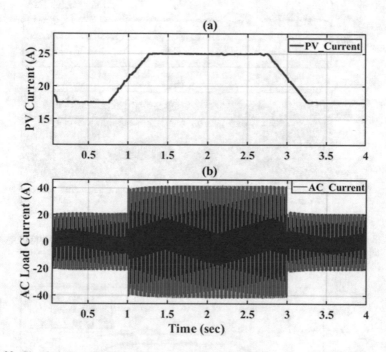

Fig. 4.22 Simulation results of the PV system with BES. (**a**) PV output current and (**b**) AC load current

sudden input of load 2 related to the model. When the load profile is changed from 3000 W to 6000 W, it leads to increase in the peak of AC current from 21.3 A to 41.85 A. Then, the AC load current peak decreases from 41.85 A to 21.3, in response to the change of the load profile from 6000 W to 3000 W, as illustrated in Fig. 4.22b. Also, the effect of change solar irradiance did not appear on the AC current due to the existence of a BES connected to the system, where the BES compensated the difference in the current to the load.

The output voltage of the PV array (V_{pv}) is maintained at 200 V as shown in Fig. 4.23a, despite the variation of solar irradiation. Therefore, the MPPT controller can accurately track the PV array voltage at the maximum power point (V_{mpp}) to harness the maximum power from the PV array during the rapid variation of solar irradiance. The proposed control on BES overcomes the transient moments resulting from the sudden switching of the load 2 and keeps up the value of the DC voltage level (output of the boost converter) at 280 V. Also, in Fig. 4.23b shows the stability of the AC voltage value. The BES compensated the voltage difference to stabilize the AC voltage on the load terminals.

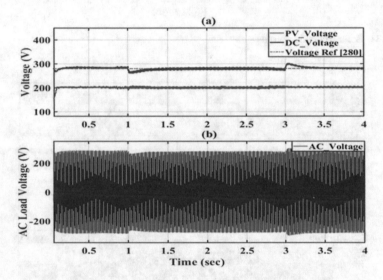

Fig. 4.23 Simulation results of the system with BES. (**a**) The PV output voltage and boost output DC voltage, (**b**) AC load voltage

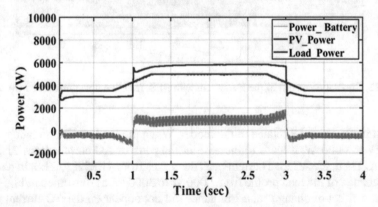

Fig. 4.24 Simulation results of the PV system without BES for the PV generated power, load power, and battery power

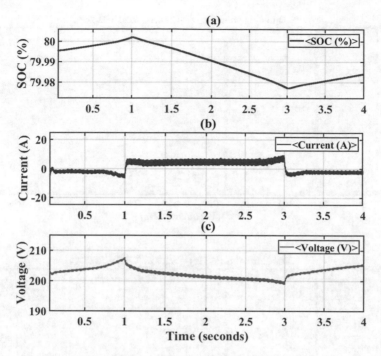

Fig. 4.25 BES response (**a**) SOC %, (**b**) battery current, and (**c**) battery voltage

In order to evaluate the validation of the MPPT technique, Fig. 4.24 shows the output power of one PV array (P_{pv}). It can be seen that the P&O MPPT technique can track accurately the MPP when the solar irradiance changes rapidly. Also, the figure shows the power of the BES and the power consumed by both loads during the system operation at the same scenario of load profile.

Figure 4.25 shows the operating curves of the battery (SOC, voltage, and current) under the effect of changing solar irradiation. In Fig 4.25a, the battery SOC varies depending on both solar radiation and load profile. In the case of an excess of power in the production of PV array, it is charged by BES. Conversely, in case of low power output from PV array, the BES will compensate for the difference in power to the load as show in Fig. 4.25b, c.

4.3.3   Voltage and Current Harmonic Analysis

The PV inverter should satisfy high power quality to meet standard recommenda tions of harmonics as dictated by national standards such as IEEE 519 and IEC 61727. The IEEE and IEC standards recommended that THD should be less than 5%, and the higher harmonic content of each individual harmonic is not more than 3% for PV system [84, 85]. The harmonic spectrum of both inverter voltage and

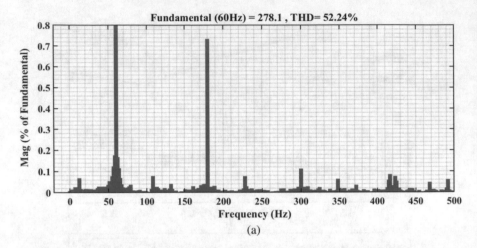

(a)

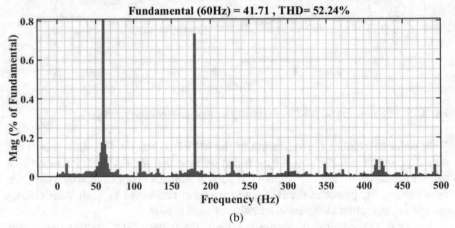

(b)

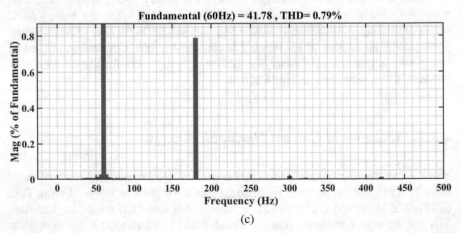

(c)

Fig. 4.26 Harmonic analysis of AC load current and AC load voltage before and after using LCL filter. (**a**) THD of the AC voltage before using LCL filter. (**b**) THD of the AC current before using LCL filter. (**c**) THD of the AC voltage after using LCL filter. (**d**) THD of the AC current after using LCL filter. (**e**) Harmonic spectrum of AC current after using LCL filter

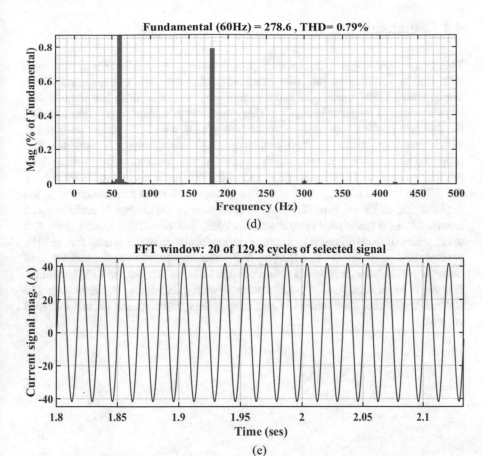

Fig. 4.26 (continued)

current of them after using LCL filter is shown in Fig. 4.26a, b. It shows the THD for both inverter voltage and current are 52.24%. The LCL filters are designed from a rating of the inverter and used to remove the harmonics which are generated from the inverter. Under steady-state operation, the voltage and current waveforms are taken to evaluate the harmonics control of the designed PV system. The standard tool in MATLAB for FFT tools is used to decide the harmonic magnitude of the AC voltage and AC current. Figure 4.26c, d shows the THD for both inverter voltage and current are 52.24% that after using LCL filter. The harmonic spectrum of AC load current after using LCL filter is shown in Fig. 4.26c.

4.4 Summary

This chapter investigated the dynamic performance of the studied PV system with BES during variation of the solar irradiance. Moreover, the effectiveness of the implemented MPPT techniques and the employed control strategy is evaluated during variations of the solar irradiance. This chapter is primarily intended to enhance the dynamic performance of the proposed PV system with BES under constant and variable solar irradiation. Furthermore, an optimal control strategy is presented. The proposed system targets in reducing the impact of transient moments resulting from the sudden entry and exit loads connected to the system. It also explains the role of the BES in how to manage system loads and demonstrates the improvement in the performance of PV systems. The system presented in the chapter is evaluated and compared to the traditional system without BES. The simulation results show that the capacity of the BES helps to improve the performance of the system through the control used in the process of loading and unloading to manage the sudden load changes and helps to maintain a stable voltage level on the load and PV terminals. It is worth to mention that the control scheme with BES ensures stable voltage and current levels and overcomes the transient spikes that appears on the AC load current.

Chapter 5
The Performance Analysis of a PV System with Battery-Supercapacitor Hybrid Energy Storage System

5.1 Introduction

In remote areas, stand-alone PV systems are most common. A typical stand-alone system incorporates a PV panel, regulator, energy storage system, and load. Generally, the most common storage technology employed is the lead-acid battery because of its low cost and wide availability. PV panels are not an ideal source for battery charging; the output is unreliable and heavily dependent on weather conditions. Therefore, an optimum charge/discharge cycle cannot be guaranteed, resulting in a low battery SOC. Low battery SOC leads to sulfation and stratification, both of which shorten battery life. Batteries are commonly implemented in stand-alone PV power systems to fulfill the power mismatch between the PV power generation and the load demand. Generally, a battery would encounter frequent deep cycles and irregular charging pattern due to the varying output of PV and the intermittent high-power demand of the load. These operations would shorten the battery life span and increase the replacement cost of the battery [86, 87].

BS-HESS is thus a practical solution to minimize the battery stress, battery size, and the total capital cost of the system. The technical characteristics of battery and SC, such as specific power, specific energy, response time, and durability, are complementary. A control strategy is essential for the BS-HESS to optimize the energy utilization and energy sustainability to a maximum extent as it is the algorithm which manages the power flow of the battery and SC. One of the common aims of BS-HESS implementation is to prolong the battery life span by reducing the peak current demand and the dynamic stress of the battery. Battery peak current reduction would reduce the internal voltage drop in the battery and improve the battery efficiency [88]. Reduction in battery dynamic stress minimizes the heating and the internal losses of the battery.

© Springer Nature Switzerland AG 2019
A. A. Elbaset et al., *Performance Analysis of Photovoltaic Systems with Energy Storage Systems*, https://doi.org/10.1007/978-3-030-20896-7_5

This chapter proposes an optimal control strategy for a stand-alone PV system with BS-HESS. The objectives of the proposed control strategy are to reduce the dynamic stress and the peak current demand of the battery while constantly considering the SOC level of the SC. The proposed control strategy comprises of a low-pass filter (LPF) and fuzzy logic controller (FLC). As the fluctuations of PV output has been taken into account in this study, the LPF filtration process is executed to allocate the high dynamic component of the power demand to the SC and refer the low dynamic component of the power demand to the FLC. The FLC is computationally efficient and it works well with optimization and adaptive techniques. Therefore, it is utilized to reduce the peak current demand of the battery by manipulating the amount of power to be charged/discharged by the SC based on the real-time power demand and the SOC level of the SC. The performance of the proposed system is compared to the conventional systems (stand-alone PV system with battery storage only, stand-alone PV system with BS-HESS with FBC) by Simulink with the setup of rural household load profile and the actual solar irradiation profile of a rainy day. The system description, modeling, and control strategy obtained results, and concluding remarks are given in the next sections.

5.2   Structure and Simulation of Stand-Alone PV Systems with BS-HESS

Figure 5.1 illustrates the simplified diagram of the stand-alone PV system with BS-HESS where the BS-HESS of the proposed system is equipped with battery unit, SC, bidirectional DC/DC converter and control circuitry. The structure and detail of PV array, battery, and SC model are available in the Simulink library. Typically, the BS-HESS takes the advantages of high-energy density storage and high-power density storage to achieve the desirable performance in which the BS-HESS is proposed in this work as shown in Fig. 5.1 [89]. However, a complex conditioning circuitry is required to combine the battery and SC as a single power source. As the SC voltage highly fluctuates due to its low-energy density, the BS-HESS of the proposed system is implemented in a semi-active topology where a bidirectional DC/DC converter is placed next to the SC to decouple the battery and SC with system. Figure 5.1 depicts the structure of semi-active BS- HESS where a power electronic unit is employed to control the power flow of the battery and SC based on the control strategy. The power electronic unit consists of a bidirectional DC/DC converter and a control circuitry. This topology allows for a sufficient degree of freedom to implement different control strategies. In addition, this topology provides a good trade-off between the performance (Fig. 5.2).

Two different models of the stand-alone PV system are constructed in MATLAB/ Simulink which are the system with battery-only system and the system with BS-HESS as illustrated in Fig. 5.3a, b. The general power equation of the system can be expressed as (5.1)

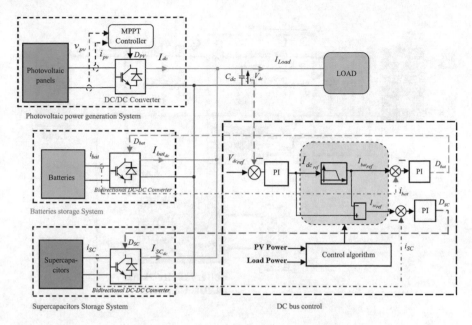

Fig. 5.1 Simplified diagram of the stand-alone PV system with energy storage

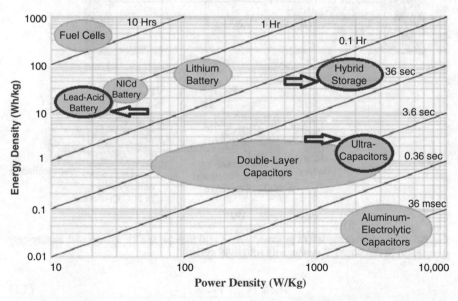

Fig. 5.2 Ragone chart showing the power density and energy density of different storages [89]. (Source: US Defence Logistics Agency)

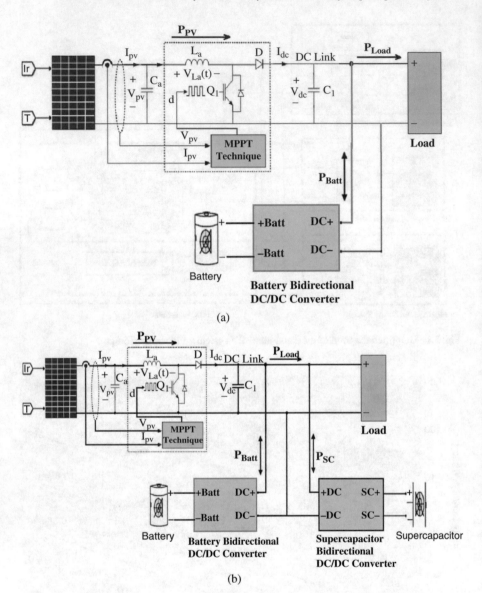

Fig. 5.3 Two different models of proposed system. (**a**) Stand-alone PV system with battery-only storage. (**b**) Stand-alone PV system with BS-HESS system

$$P_{PV} \mp P_{Batt} \mp P_{SC} = P_{Load} \tag{5.1}$$

where P_{PV} is the power generation of PV array, P_{Batt} is the power flow of BES system, P_{SC} is the power flow of SC storage system, and P_{Load} is the power demand of the load.

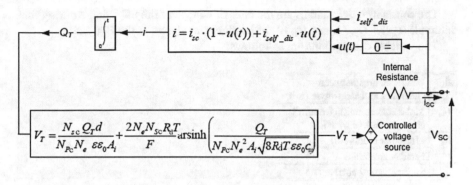

Fig. 5.4   Supercapacitor equivalent circuit model [92]

5.2.1   Supercapacitor Model

The SC block implements a generic model parameterized to represent the most popular types of SC. Figure 5.4 shows the equivalent circuit model for the SC. The model consists of two components, the controlled voltage source, and the equivalent internal series resistance. The internal series resistance is a loss term that models the internal heating in the capacitor and is most important during charging and discharging. Also, it models the current leakage effect and will impact the long-term energy storage performance of the SC [90]. Equation (5.2) describes the SC output voltage (V_{SC}) using a Stern equation [91].

$$V_{SC} = \frac{N_{sc} Q_T d}{N_{pc} N_e \varepsilon \varepsilon_0 A_i} + \frac{2 N_e N_{sc} R_d T}{F} \sinh^{-1}\left(\frac{Q_T}{N_{pc} N_e^2 A_i \sqrt{8 R_d T \varepsilon \varepsilon_0 C_m}} \right) - R_{SC} \cdot i_{SC} \quad (5.2)$$

$$\text{With}: Q_T = \int i_{SC}\, dt \quad (5.3)$$

To represent the self-discharge phenomenon, the SC electric charge is modified as follows (when $i_{SC} = 0$):

$$Q_T = \int i_{\text{self\_dis}}\, dt \quad (5.4)$$

where

$$i_{\text{self\_dis}} = \begin{cases} \dfrac{C_T \alpha_1}{1 + s R_{SC} C_T} & \text{if } t - t_{oc} \leq t_3 \\[2ex] \dfrac{C_T \alpha_2}{1 + s R_{SC} C_T} & \text{if } t_3 < t - t_{OC} \leq t_4 \\[2ex] \dfrac{C_T \alpha_3}{1 + s R_{SC} C_T} & \text{if } t - t_{oc} \geq t_4 \end{cases} \quad (5.5)$$

The constants α_1, α_2, and α_3 are the rates of change of the SC voltage during time intervals (t_{oc}, t_3), (t_3, t_4), and (t_4, t_5), respectively, as shown in the Fig. 5.5:

The variable descriptions are as follows:

A_i	Interfacial area between electrodes and electrolyte (m²)
C_m	Molar concentration (mol/m³) equal to $c = 1/(8\,N_A\,r^3)$
R_d	Molecular radius (m)
F	Faraday constant
i_{SC}	Supercapacitor current (A)
V_{SC}	Supercapacitor voltage (V)
C_T	Total capacitance (F)
R_{SC}	Total resistance (ohms)
N_e	Number of layers of electrodes
N_A	Avogadro constant
N_{pc}	Number of parallel SCs
N_{sc}	Number of series SCs
Q_T	Electric charge (C)
R	Ideal gas constant
D	Molecular radius
T	Operating temperature (K)
ε	Permittivity of material
ε_0	Permittivity of free space

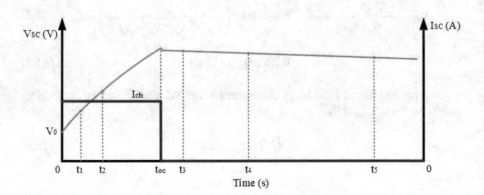

Fig. 5.5 Time intervals of charge and self-discharge characteristic for SC

5.2.2   *Control Circuit of Bidirectional DC/DC Buck/Boost Converter*

In PV systems, bidirectional DC/DC converters are employed to link up the high-voltage DC bus to the BS-HESS. Here they are needed to regulate the power supply for electrical loads to help them provide their power demanded. The bidirectional DC/DC converters can be classified into two categories depending on the galvanic isolation between the input and output side. In this book, the non-isolated bidirectional half-bridge DC/DC converters is used to connect the battery and SC with the PV system, which illustrated its circuit diagram in Fig. 5.6. The circuit was explained in Chap. 4.

Figure 5.7 shows the control circuit for controlling the work of the bidirectional DC/DC converter. The control circuit generates the control signal to the gate of switches Q1 and Q2. The gate signal switches on the function of the bidirectional DC/DC converter between the boost and the buck depending on the power control signal.

5.3   Control Strategies of HESS

The control strategy manages the power flow of the HESS based on the real-time system conditions. It is usually complex and required to operate continuously in order to fulfill the multiple objectives. Optimal control of the HESS is crucial to optimize the energy utilization and sustainability to a maximum extent. The common aims of the control strategies are listed as follows:

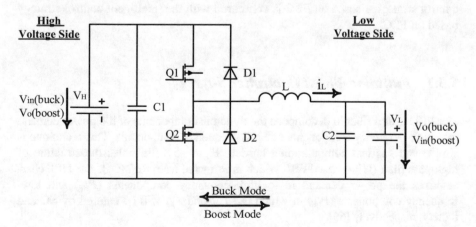

Fig. 5.6 Circuit diagram of half-bridge bidirectional DC/DC converter

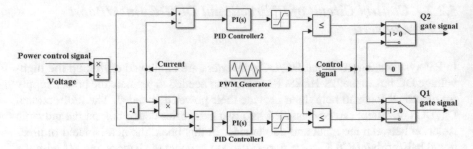

Fig. 5.7 Control circuit of bidirectional DC/DC converter

- To reduce the peak power demand, charge/discharging cycle, and dynamic stress level of the battery
- To prevent the deep discharge of the battery
- To maintain a stable DC voltage
- To reduce the loss of power supply possibility and operational and maintenance cost
- To improve the overall efficiency of the system

Generally, the control strategies can be characterized as classical control strategies and intelligent control strategies. The classical control strategies such as FBC are simple and easy to be implemented as they do not require complicated processing. However, they are normally sensitive to the parameter variation and rigid. Intelligent control strategy such as fuzzy logic controller (FLC) is more robust and efficient compared to classical control strategies as it enhances the dynamic behavior of the system without requiring an exact model of the system. However, the MFs of FLC are usually determined by using the trial-and-error method which is time-consuming and lacking optimization [93]. In this chapter, the classical control strategies based on FBC is compared with the intelligent control strategy based on FLC.

5.3.1   Filtration-Based Controller Strategy

The FBC uses a filter to decompose the dynamic components of the power demand into high-frequency components and low-frequency components. This technique is simple and has less computational burden. Figure 5.8 illustrates the structure of high-pass filter (HPF)-based FBC which is extracted from Ref. [94]. The HPF characterizes the power demand in to high-frequency components (P_{HF}) and low-frequency components (P_{LF}) in which the P_{HF} and P_{LF} will be catered by SC and battery, respectively [63].

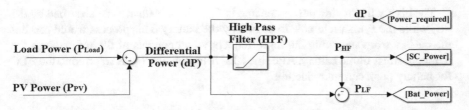

Fig. 5.8 Structures of the filtration-based controller based on FBC

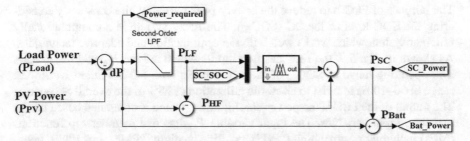

Fig. 5.9 Structures of the intelligent control strategy based on LPF and FLC

5.3.2   Intelligent Control Strategy Based on LPF and FLC

The structure of the intelligent control strategy is illustrated in Fig. 5.9, which aims
to minimize the dynamic stress and the peak current demand of the battery. The
control strategy comprises of two parts that are the LPF and FLC. The structure of
the proposed control strategy is explained in the following sections.

5.3.2.1   Low-Pass Filter (LPF)

The generation power from PV and the demand power for load are highly fluctuating. In the conventional system, the battery is stressed to satisfy the highly fluctuating (dP). The highly fluctuating battery current would produce an extensive heat
inside the battery which leads to an increased battery internal resistance and lower
efficiency.

Therefore, LPF is implemented to reduce the dynamic stress of the battery by
decomposing the dP into P_{HF} and P_{LF}. The P_{LF} is the output of LPF, while the P_{HF} is
the difference between dP and P_{LF} [95].

$$P_{LF} = \text{Lowpass filter} (dP) \tag{5.6}$$

$$P_{HF} = dP - P_{LF} \tag{5.7}$$

The highly fluctuating power demand is P_{HF} which is ideal to be absorbed by the SC, while the P_{LF} is preferable to be met by the battery. This process would prevent the battery from supplying the high-frequency components of dP and reduces the dynamic stress of the battery. After the LPF filtration, the P_{LF} is referred to the FLC for battery peak current reduction.

5.3.2.2   Fuzzy Logic Controller (FLC)

The purpose of FLC is to reduce the battery peak current while constantly considering the SOC level of the SC (SOC_{SC}). The fuzzy system is a computationally efficient system which works well with the optimization and adaptive techniques. As shown in Fig. 5.9, the FLC has two inputs which are the P_{LF} and the SOC_{SC}. The operating range of SOC_{SC} of the two models with SC is limited within the range of 50–100% in order to allow the utilization of 75% of the overall SC energy. The output of the FLC is the power-sharing ratio, which is computed based on the real-time input variables. The input variable P_{LF} has five membership functions (MFs) including positive high ("PH"), positive medium ("PM"), low ("L"), negative low ("NL"), and negative high ("NH") as shown in Fig. 5.10a. The positive P_{LF} is the power demand to be supplied by the HESS, and the negative P_{LF} is the excessive power to be absorbed by the HESS. On the other hand, the input variable SOC_{SC} has only three MFs, namely, high ("H"), medium ("M"), and low ("Low"), as shown in Fig. 5.10b. Meanwhile, the output variable a has five MFs that are PH, PL, zero ("Z"), NL, and NH as shown in Fig. 5.10c. The positive and negative of MFs indicate the power ratio to be supplied and absorbed by the SC, respectively.

The rules of the FLC are listed in Table 5.1. When the power demand of the P_{LF} is "L," the power-sharing ratio would be "Z" regardless of the SOC_{SC} condition as the low power demand imposes little stress to the battery. When the P_{LF} is positive, a is set according to the level of the power demand and the SOC_{SC} in order to reduce the peak current demand of the battery. When the P_{LF} is negative, a is set based on the excessive power and the SOC_{SC} level to recover the charge of the SC.

The total power to be shared by the SC (P_{SC}) can be calculated by using Eq. (5.8):

$$P_{SC} = \beta + P_{HF} \tag{5.8}$$

where β is the output signal from FLC and the battery is expected to supply the power mismatch between P_{SC} and dP as defined in Eq. (5.9).

$$P_{Batt} = dP - P_{SC} \tag{5.9}$$

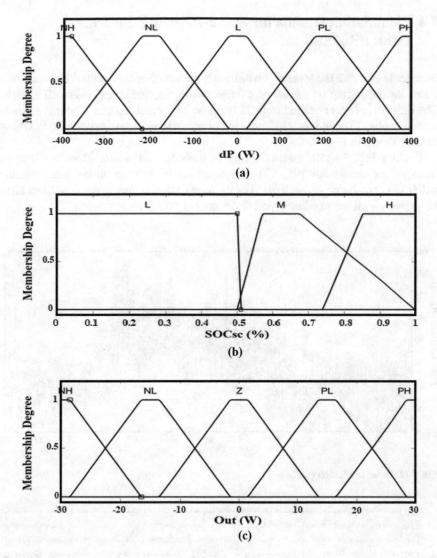

Fig. 5.10 Inputs and output membership functions for FLC. (**a**) Input 1: P_{LF}. (**b**) Input 2: SOC$_{SC}$. (**c**) Output MFs

Table 5.1 Fuzzy logic rules

Rules of FLC		Input 1: dP				
		PH	PL	L	NL	NH
Input 2: SOC$_{sc}$	H	PH	PL	Z	Z	Z
	M	PL	PL	Z	NL	NL
	L	Z	Z	Z	NL	NL

5.4   Simulation Results for Stand-Alone PV Systems with BS-HESS

In order to evaluate the system performance, the actual solar irradiation profile of a rainy day depending on the set of references [63, 95, 96] is applied in all models. The time period was changed from 24 hours to 24 seconds so that 1 hour is actually represented to 1 second on the simulation program to reduce the simulation running time as shown in Fig. 5.11.

Besides, Fig. 5.12 illustrates the rural household load profile which is extracted from set of references [95, 97] and modified to impose more stress on the BS-HESS. The high-power demand of the load occurred during the time from 12 to 18 seconds with the maximum power demand of 403.5 W.

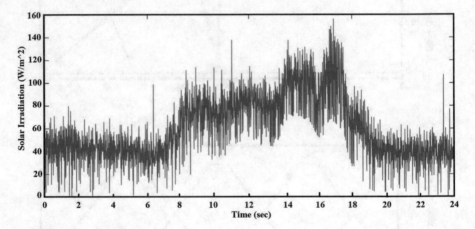

Fig. 5.11   Solar irradiation profile

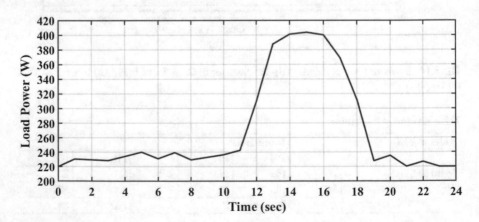

Fig. 5.12   Profile of load demand

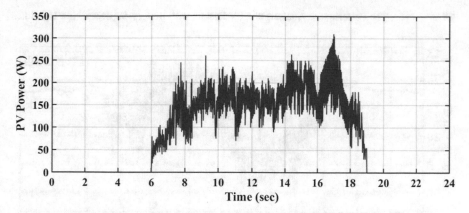

Fig. 5.13 Output power from PV array

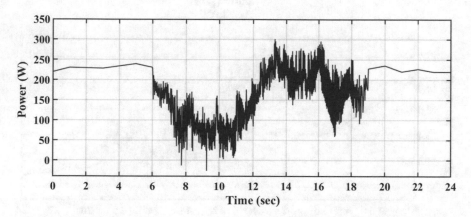

Fig. 5.14 Power mismatch between PV power generation and load demand power (dP)

Table 5.2 The configuration of the models in Simulink

Model No.	Energy storage system	Control strategy
Model 1	Battery	–
Model 2	Battery + SC	FBC
Model 3	Battery + SC	LPF with FLC

Table 5.3 The specification of the stand-alone PV system with BS-HESS

Component	Rating	
PV array	Power	1200 W
Battery	Type	Lead-acid
	Voltage	12 V
	Capacity	7 Ah
SC	Voltage	16 V
	Capacitance	58 F

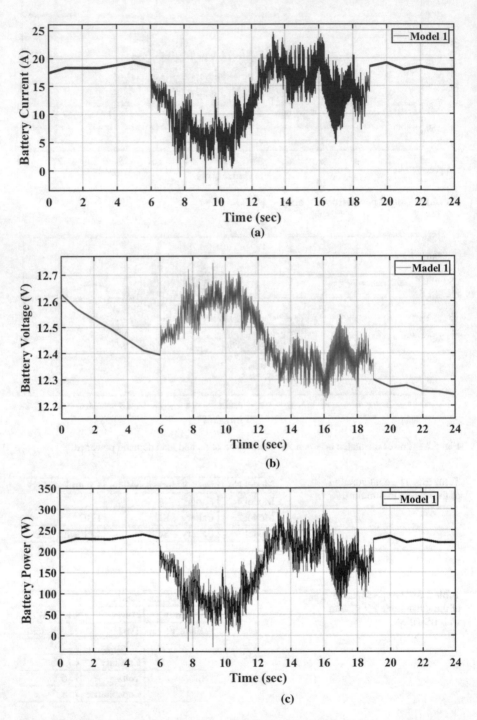

Fig. 5.15 Battery current, voltage, power, and SOC (%) for Model 1. (**a**) Battery current (A) for Model 1. (**b**) Battery voltage (V) for Model 1. (**c**) Battery power (W) For Model 1. (**d**) Battery SOC (%) for Model 1

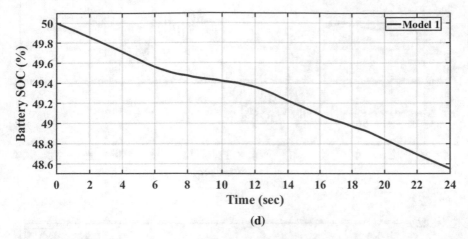

Fig. 5.15   (continued)

Figure 5.13 demonstrates the power generation of PV based on the solar irradiation profile in Fig. 5.11. By referring to Eq. (5.10), the power demand of the BS-HESS, which is the power deficit between PV output power and the load demand, is illustrated in Fig. 5.14.

$$dP = P_{PV} - P_{Load} = P_{Batt} + P_{SC} \qquad (5.10)$$

The three different models of the stand-alone PV system were constructed in MATLAB/Simulink as listed in Table 5.2. Model 1 is the conventional stand-alone PV system with a battery-only system. Model 2 is the stand-alone PV system with BS-HESS using classical control strategies based on HPF. Model 3 is the stand-alone PV system with BS-HESS using intelligent control strategies based on LPF and FLC. The specification of the system main component (PV array, battery, and SC) is listed in Table 5.3.

Several battery parameters particularly battery peak current (I_{batt_peak}), battery peak power (P_{batt_peak}), average battery SOC (SOC_{batt_avarge}), and final battery SOC (SOC_{batt_final}) are evaluated. The reduction of I_{batt_peak} and P_{batt_peak} would lead to lower battery stress, higher battery efficiency, and reduction of internal voltage in the battery. The SOC_{batt_avarge} and SOC_{batt_final} are evaluated in this study in which higher SOC_{batt_avarge} and SOC_{batt_final} would extend the battery lifetime and reduce the loss of power supply possibility of the system.

For Model 1, the battery is the only energy storage system to satisfy the power mismatch between the PV output power and the load demand. Figure 5.15 shows the operating curves of battery for Model 1 in the simulation. In Fig. 5.15a, it is illustrated that the battery current has high fluctuation. Hence, the battery voltage as shown in Fig. 5.15b is still highly fluctuating. As a result, the battery power profile

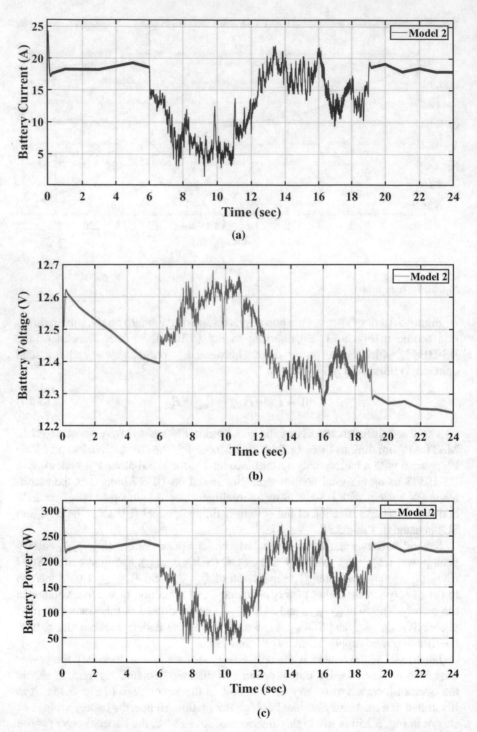

Fig. 5.16 Battery current, voltage, power, and SOC (%) for Model 2. (**a**) Battery current (A) for Model 2. (**b**) Battery voltage (V) for Model 2. (**c**) Battery power (W) For Model 2. (**d**) Battery SOC (%) for Model 2

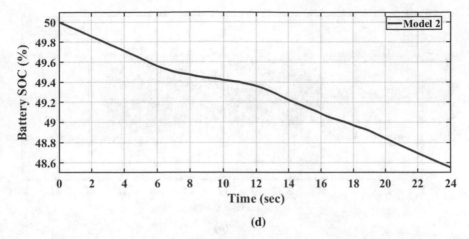

(d)

Fig. 5.16 (continued)

is identical to the profile of dP as shown in Fig. 5.14 where the battery always experiences highly fluctuating discharging current.

For Model 2, Fig. 5.16 shows the operating curves of the battery from which we observe that the dynamic stress level of the battery has a considerable reduction, but the I_{batt_peak} Figure 5.16a and P_{batt_peak} Figure 5.16c are not improved significantly. This is because of the FBC that is designed to reduce the dynamic stress of the battery without considering the peak demand. The dynamic stress of the battery is slightly improved as compared to the battery-only system. The dynamic stress level of battery voltage has reduction as shown in Fig. 5.16b.

On the other hand, the SOC_{batt_avarge} and SOC_{batt_final} are not improved substantially (~0%) as shown in Fig. 5.15d; only the highly fluctuating low power components are absorbed by the SC in Fig. 5.17. Figure 5.17c illustrates that the SC absorbs/supplies only a small portion of dynamic components of the power demand. The SC current, voltage, and the SOC_{SC} of Models 2 throughout the simulation are illustrated in Fig. 5.17a–c, respectively.

For Model 3, a system with intelligent control (LPF with FLC), Figs. 5.18a, b illustrate that the battery current and voltage according to the predefined rules and battery peak current are reduced. The system has improved final SOC and average SOC of the battery as illustrated in Fig. 5.18d. It is evident that the battery power profile is smoother than Model 1 and 2 as shown in Fig. 5.18c.

The mismatch between dP and battery power is compensated by SC as shown in Fig. 5.18. The SC is charging and discharging at a high frequency as it absorbs the fast-transient component of the power demand. The P_{HF} is a highly fluctuating power demand which is desirable to be supplied by the SC for reduce the battery peak demand (Fig. 5.19).

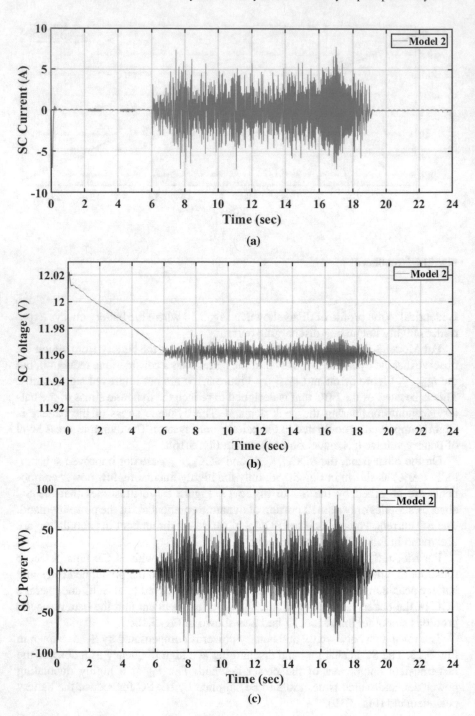

Fig. 5.17 Operating curves of the SC for Model 2. (**a**) SC current (A) for Model 2. (**b**) SC voltage (V) for Model 2. (**c**) SC power (W) For Model 2. (**d**) SC-SOC (%) for Model 2

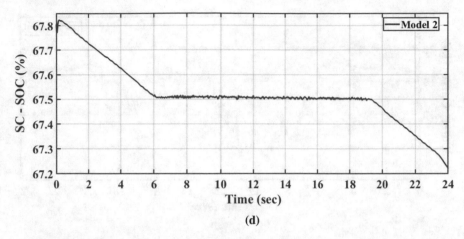

(d)

Fig. 5.17   (continued)

Table 5.4 summarizes and compares the battery and SC performance of all the models. Model 3 has improved final SOC and average SOC of the battery by (0.1133%) and (0.226%), respectively. Table 5.4 shows that it has the best performance in terms of peak current reduction (18.031%) and the reduction of battery deep discharge as the strategy is designed to minimize the peak power demand of the battery. Meanwhile the SC discharges appropriately to meet the peak demand by constantly considering the SOC level of SC. Hence, the I_{batt_peak} and P_{batt_peak} are reduced by 18.031% and 18.098%, respectively, in comparison to Model 1.

Figure 5.20 shows a comparison of the battery current in all models with a focus on the difference between them at some operating times.

In addition, a comparison of the battery voltage in all models is performed and presented in Fig. 5.21, in order to show the differences between the control methods used with each model.

The most important parameters on the efficiency of the system is to control the battery power and reduce its peak and reduce the fluctuations; Fig. 5.22 shows the comparison of battery power in all models.

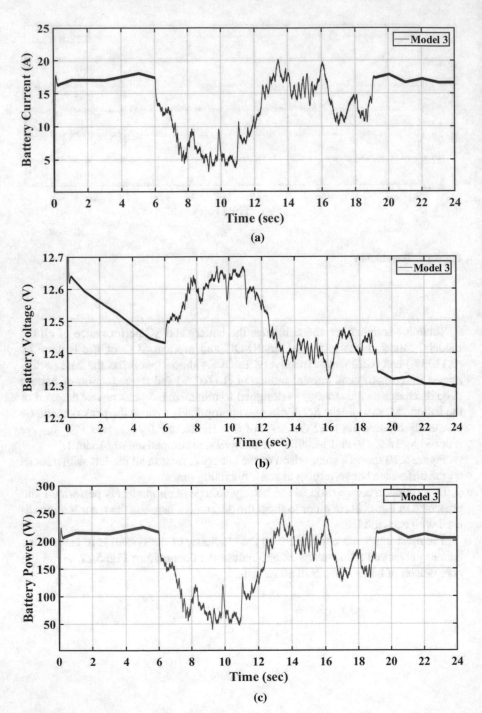

Fig. 5.18 Battery current, voltage, power, and SOC (%) for Model 3. (**a**) Battery current (A) for Model 3. (**b**) Battery voltage (V) for Model 3. (**c**) Battery power (W) For Model 3. (**d**) Battery SOC (%) for Model 3

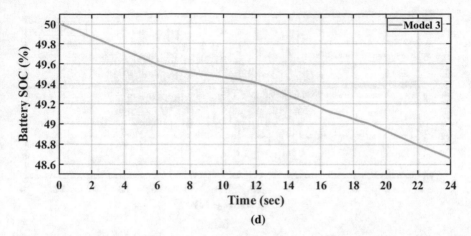

(d)

Fig. 5.18 (continued)

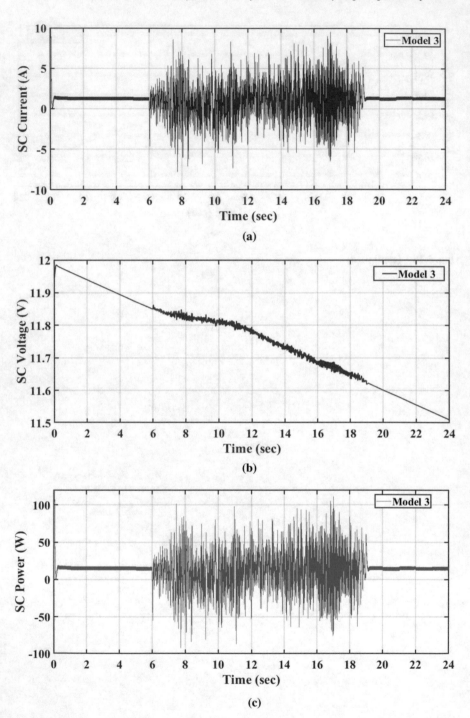

Fig. 5.19 SC current, voltage, power, and SOC (%) for Model 3. (**a**) SC current (A) for Model 3. (**b**) SC voltage (V) for Model 3. (**c**) SC power (W) For Model 3. (**d**) SC- SOC (%) for Model 3

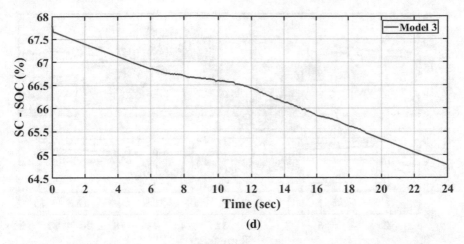

(d)

Fig. 5.19 (continued)

Table 5.4 Summary and comparison of the battery and SC performance of all the models

Parameters	unit	Model 1	Model 2	Model 3
I_{batt_peak}	Current (A)	24.68	22.09	20.23
	Reduction (%)	–	10.494	18.031
P_{batt_peak}	Power (W)	302.8	271.8	248
	Reduction (%)	–	10.24	18.098
SOC_{batt_final}	SOC (%)	48.5519	48.5521	48.6617
	Increment (%)	–	0.0004	0.226
SOC_{batt_avarge}	SOC (%)	49.275	49.2761	49.3309
	Increment (%)	–	0.0022	0.1133
I_{SC_peak}	Current (A)	–	7.78	9.547
	Increment (%)	–	–	18.508
P_{SC_peak}	Power (W)	–	91.32	110.6
	Increment (%)	–	–	17.432
SOC_{SC_final}	SOC (%)	–	67.2218	64.7838
	Reduction (%)	–	–	3.626

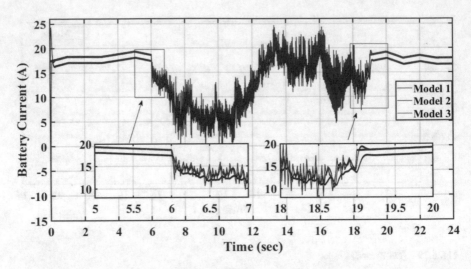

Fig. 5.20 Comparison of the battery current of all the models

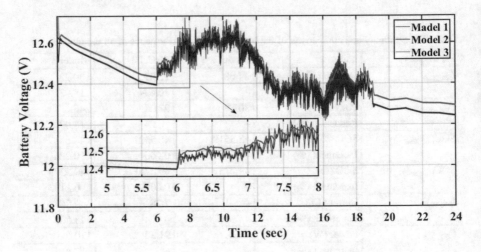

Fig. 5.21 Comparison of the battery voltage of all the models

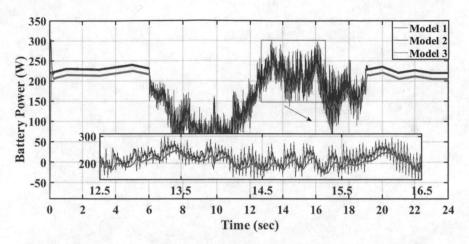

Fig. 5.22 Comparison of the battery power of all the models

5.5   Summary

In this chapter, the Simulink model of the proposed stand-alone PV system with BS-HESS (Model 3) and an optimal control strategy are presented. The objectives of the system are to reduce the dynamic stress and peak power demand of the battery by employing LPF and FLC. The FLC is used order to optimize the battery peak current reduction. The proposed system is evaluated and compared to the conventional system with battery-only systems and the systems with classic control strategies (FBC). The BS-HESS shows the positive impact to the battery and the overall system. The simulation results show that the dynamic stress and peak current demand of the battery in the proposed system are greatly improved, which will eventually extend the battery life span. The proposed system is able to operate the SC within the recommended SOC range and utilize the limited energy of SC effectively to perform better than the conventional systems.

Chapter 6
Experimental Work

6.1 Introduction

Energy storage systems used in renewable energy system for storing the energy when renewable power is generated and releasing power when renewable power is not sufficient. BESs provide immediate energy storage in a rechargeable battery. High-performance batteries and battery chargers are necessary for high-efficiency, fast-response, high-power, and high-energy density. In this chapter, the experimental setup along with its components is implemented in renewable energy laboratory, Faculty of Industrial Education, Suez University, Suez, Egypt. This chapter includes two parts; the first part presents the experimental setup of an off-grid PV system, and the second part contains the experimental results and discussion. The experimental study focuses on the effects of using BES with variable irradiation and load profile on the off-grid PV system.

6.2 Experimental Setup of Off-Grid PV System

The main purpose of this research is to develop and investigate the performance of the PV stand-alone systems with energy storage systems. Figure 6.1 illustrates an experimental setup of stand-alone PV system with BES and different loads. In this figure, the PV array is consisting of three modules which are connected to solar charge controller MPPT; the DC/DC converter is controlled with MPPT to maximize the output power with the appropriate output voltage and current of the PV. The battery charger is responsible for charging the battery when the PV is providing power. When the renewable power is not sufficient, the battery releases power to the DC/DC converter for providing a DC link voltage of the DC/AC off-grid inverter. The inverter converter injects the AC power to the AC loads. The components specifications of the experimental setup are listed in Table 6.1.

© Springer Nature Switzerland AG 2019   101
A. A. Elbaset et al., *Performance Analysis of Photovoltaic Systems with Energy
Storage Systems*, https://doi.org/10.1007/978-3-030-20896-7_6

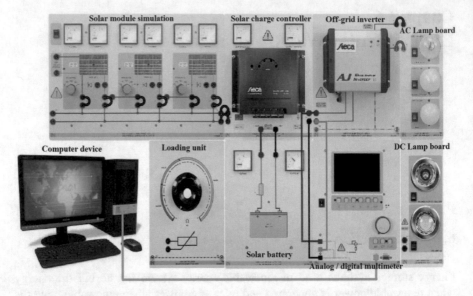

Fig. 6.1 Experimental setup for stand-alone PV system with AC load

Table 6.1 The specification of the experimental stand-alone PV system elements

Component	Rating	
PV modules	Power	3×38 W
Solar battery	Type	Lead acid
	Voltage	12 V
	Capacity	12 Ah
Off-grid inverter	Voltage	12 V/230 V
	Frequency	50 Hz
Load unit	Power	500 W
Solar charge controller-MPPT	Voltage	12–24 V
	Current	10 A
DC Lamp board	Voltage	12 V
AC Lamp board	Voltage	230 V

6.2.1   Elements of the Experimental Setup

6.2.1.1   Solar Modules Simulation

Training panel CO3208-1A titled "Solar modules simulation" comprises three independent solar modules as shown in Fig. 6.2. The electrical specifications are as follows:

- Three separate solar modules.
- Each solar module has an adjustable irradiance.

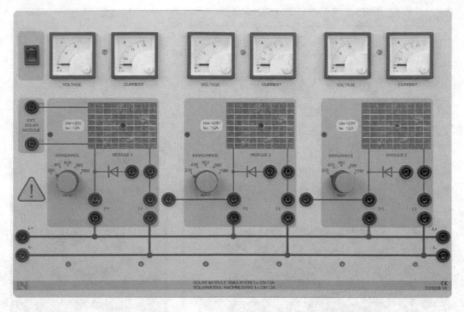

Fig. 6.2 Solar modules simulation

- Each solar module's outputs are protected against short circuit.
- Open-circuit voltage: approximately 23 V
- Short-circuit current: Up to 2A
- Integrated displays of voltage and current
- Operating voltage: 100–240 V and 50/60 Hz

The PV array is composed of three modules that are connected in parallel as shown in Fig. 6.3. For a constant temperature and different solar irradiations (200:1000 W/m²), the I-V and P-V characteristics of the three PV modules are shown in Figs. 6.4 and 6.5. From Fig. 6.5, it can be easily realized that as the solar irradiation increases, the maximum power generation increases. Similarly, in Fig. 6.4, it is observed that as the solar irradiation increases, the PV module output current increases.

6.2.1.2   Solar Charge Controller-MPPT

Training panel CO3208-1 M titled "Solar charge controller" can be used to charge lead-acid batteries using solar power as illustrated in Fig. 6.6. The electrical speci fications are as follows:

- Colored LEDs for indicating operating states.
- Integrated displays for the load connection's voltage and current.
- Deep-discharge protection for the connected battery.

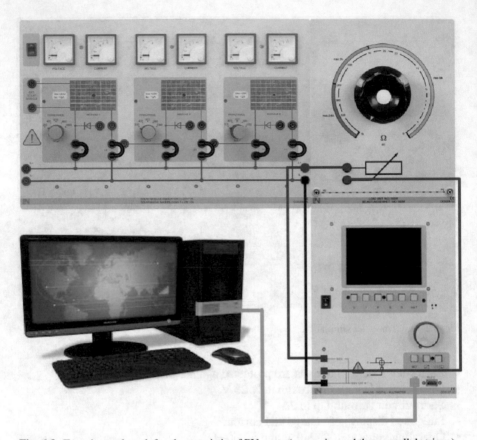

Fig. 6.3 Experimental work for characteristic of PV array (one series and three parallel strings)

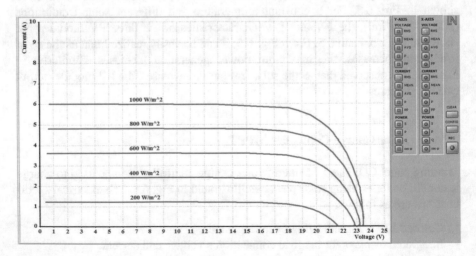

Fig. 6.4 I-V curves of PV array with different solar irradiations

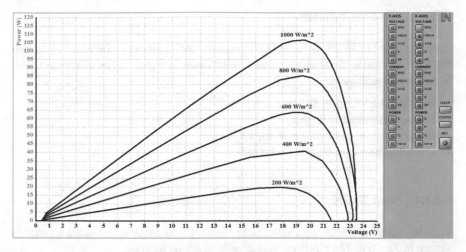

Fig. 6.5   P-V curves of PV array with different solar irradiations

Fig. 6.6   Solar charge
controller MPPT

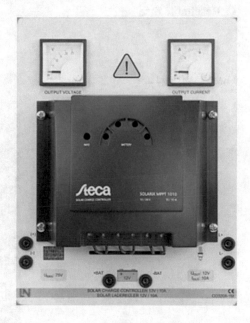

- Overload protection for the connected battery.
- Connect the battery to the charge controller. During start-up, always connect the battery before the solar generator.
- Connect the PV generator to the charge controller. The generator's voltage must not exceed 75 V.
- Connect the load to the charge controller.

6.2.1.3   Off-Grid Inverter

Training panel CO3208-1F titled "Off-grid inverter" contains an inverter for operating a PV system in stand-alone mode as illustrated in Fig. 6.7. The electrical specifications are as follows:

- Inverter (12 V/230 V, 50 Hz)
- Sinusoidal output voltage
- Reverse polarity protection on the DC side
- Deep-discharge protection for batteries

6.2.1.4   Solar Battery

Training panel CO3208-1E titled "Solar battery" contains a lead-acid battery as shown in Fig. 6.8. The electrical specifications are as follows:

- Maintenance-free lead-acid battery (12 V/12 Ah)
- Integrated displays for voltage and current
- Resettable fuse

Fig. 6.7 Off-grid inverter

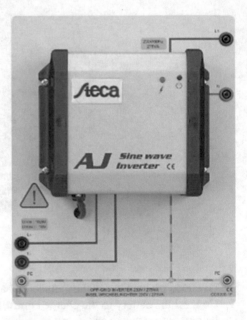

Fig. 6.8 Lead-acid battery

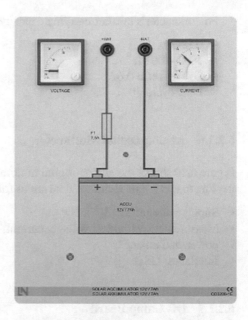

Fig. 6.9 Load
unit – 500 W

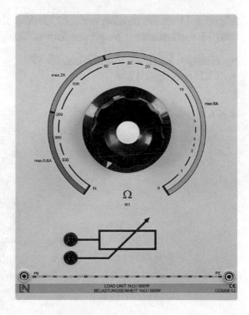

6.2.1.5 Load Unit: 500 W

Figure 6.9 shows Training panel CO3208-1 J titled "Load unit – 500 W" is used to set various operating points and record characteristics. The electrical specifications are as follows:

- Potentiometer's current-carrying capacity

  - $0\,\Omega$–$30\,\Omega$: 6A
  - $30\,\Omega$–$200\,\Omega$: 2A
  - $200\,\Omega$–$1\mathrm{k}\Omega$: 0.6A
  - Duty cycle 40%

6.2.1.6   Analog-Digital Multimeter

Figure 6.10 shows the analog-digital multimeter possesses USB interfaces for connecting to a PC. The technical data are as follows:

- Supply voltage: 230 V/50 Hz
- Measurement variables: voltage, current, active power, apparent power, reactive power, and cosine φ
- Interfaces: USB

6.2.1.7   DC Lamp Board

Figure 6.11 shows training panel CO3208-1 K titled "Lamp board – 12V" incorporates two 12 V consumers comprising lamps. The electrical specifications are follows:

- Halogen lamp 12 V and Max. 25 W
- LED spotlight 12 V

Fig. 6.10 Analog-digital multimeter

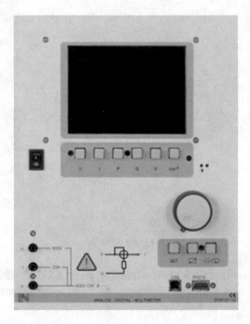

Fig. 6.11 DC lamp board

6.2.1.8   AC Lamp Board

Figure 6.12 illustrated training panel CO3208-1 L that provides three 230 V consumers in the form of lamps. The electrical specifications are as follows:

- LED lamp 230 V and 9 W
- LED lamp 230 V and 6 W
- Energy-saving lamp 230 V and 10 W

6.3   Experimental Results and Discussion

In this section, the experimental results are presented, where the results are divided into three parts as follows: Model 1(without battery)results of the operation of the system without a battery connected with it; Model 2(with battery) results of the operation of the system with a battery connected with it and clarification of the difference between the operating results of the system with and without battery. In the two previous operating cases, the results are obtained with the installation of solar irradiation at 1000 W/m^2; in Model 3 (with battery and changing solar irradiation), the operating results of the system connected to the battery are explained with the change of the solar irradiation and the electrical loads.

Fig. 6.12 AC lamp board

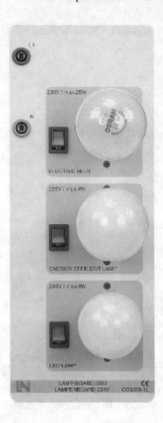

6.3.1 Model 1: Experimental Results of the System Without Battery

The PV array is composed of three modules that are connected in parallel as shown in Fig. 6.3. The following Table 6.2 presents the electrical specification of each module. The operation of Model 1 is based on the three electrical loads (10 W, 6 W, and 9 W) that are connected to the PV system in an increasing manner as shown in Fig. 6.13. In this subsection, the operating results of experimental work for the PV system without a connected battery are reviewed. The solar irradiation value applied to the three models is fixed at (1000 W/m²). The three PV modules are connected together in parallel to have a maximum output power of 106 W.

The operating results of experimental work for Model 1 are shown in the following figures where the current, voltage, and power values are displayed for the output of the PV modules, solar charge controller, and off-grid inverter. These results were recorded using a single measuring device with a change in its terminals to the places to be measured, so there are some slight differences in the entry and exit of loads due to repeated experiment three times in the same scenario.

Figure 6.14 illustrates the results of Model 1. The PV output of voltage (V_{PV}), current (I_{PV}), and power (P_{PV}) are shown in Fig. 6.14, where the change of loads

Table 6.2 Electrical specification of PV modules

Parameters	Symbol	Value
Rated power for one module	$P_{PV-module}$	35.2 W
Short-circuit current	I_{OC}	2 A
Open-circuit voltage	V_{SC}	23 V
Maximum power for array	P_{max}	106 W
Voltage at maximum power point	V_{mpp}	19.6 V
Current at maximum power point	I_{mpp}	1.8 A

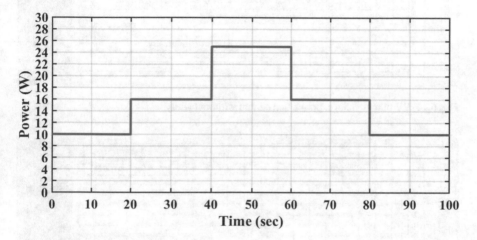

Fig. 6.13 Variable load profile for Model 1

profile is an effect on PV voltage and its value decreases as the load increases. Changes in the PV output voltage did not appear on the output voltage of the charger controller (V_{dc}), to remain stable at 14 V as shown in Fig. 6.15, thus maximizing the required load power (P_{Load}), of only 25 W. Figure 6.16 shows the voltage(V_{ac-rms}), current (I_{ac-rms}), and output power ($S - P - Q$) of the off-grid inverter or the terminals of the electric AC loads. The DC voltage from charge controller was converted from 14 V_{dc} to 225 V_{ac-rms} using the inverter to suit the electrical AC loads.

6.3.2 Model2: Experimental Results of the System with Battery

The operation of Model 2 is based on three electrical AC loads (10 W, 6 W, and 9 W) and one electrical DC load (25 W) that are connected to the PV system in an increasing manner as shown in Fig. 6.17. The difference is that in Model 2 the battery is connected to the system for study its effect on the whole system and to explain differences from Model 1. Figure 6.18 illustrates the results of Model 1. The PV output of voltage (V_{PV}), current (I_{PV}), and power (P_{PV}) is shown in Fig. 6.18. The power generated from the PV modules is greater than the power of the loads demand

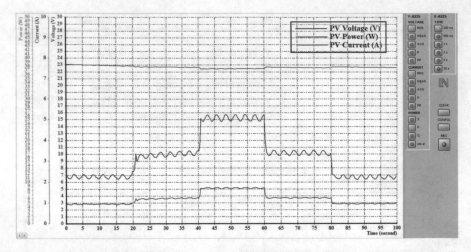

Fig. 6.14 PV output voltage, power, and current without battery

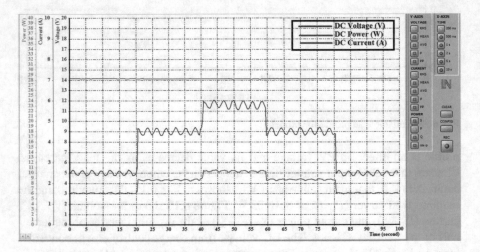

Fig. 6.15 Solar charge controller output voltage, power, and current without battery

because there is a battery connected to the system and is in charge. In Fig. 6.19, changes in DC current (I_{dc}) and DC power (P_{dc}) are consistent with loads changes. Figure 6.20 shows the voltage (V_{ac-rms}), current (I_{ac-rms}), and output power ($S - P - Q$) of the off-grid inverter. When using a battery, it has improved the active power of load as well as reduced the fluctuations in the voltage and current.

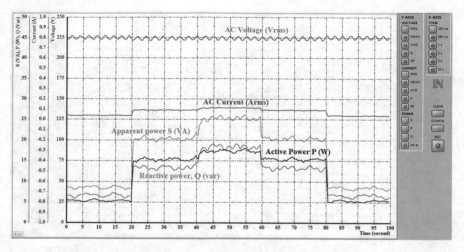

Fig. 6.16 Inverter output voltage, power, and current without battery

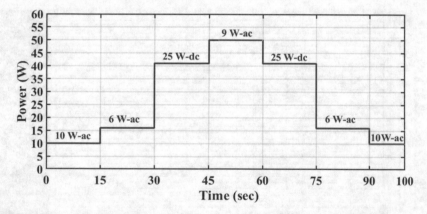

Fig. 6.17 Variable load profile for Model 2

6.3.3 Model 3: Experimental Results of the System Connected to Battery and Changing Solar Irradiation

Model 3 studies the effect of batteries on PV systems in the case of changes in loads profile and solar irradiation that applied to PV modules. Figure 6.21 shows the change in solar irradiation (from 200 to 1000 W/m²) where it increases by 200 W/m² every 15 seconds, until the irradiation reaches 1000 W/m² and drops back to 200 W/m² in the same time periods. The operation of Model 3 is based on the three electrical AC loads (10 W, 6 W, and 9 W) that are connected to the PV system in an increasing manner as shown in Fig. 6.22.

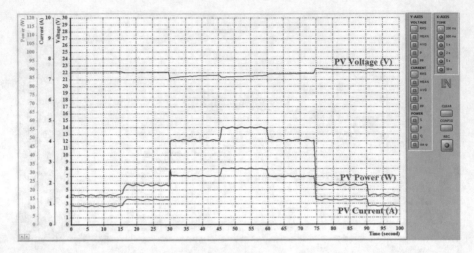

Fig. 6.18 PV output voltage, power, and current with battery

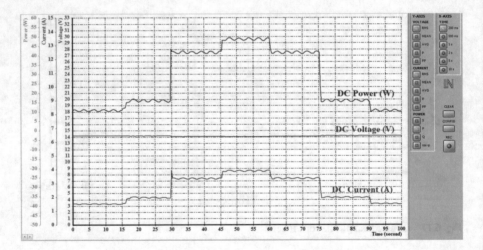

Fig. 6.19 Solar charge controller output voltage, power, and current with battery

Figure 6.23 shows the experimental results of Model 3, which consists of a PV system with a battery connected to it. The results of the experiment are obtained when the solar irradiation was changed and showed an effect on output voltage from PV as shown in Fig. 6.23. The PV output current (I_{PV}), and power (P_{PV}) are changed according to the change of the load profile from entry and exit loads. Figure 6.24 illustrates the output signal from solar charge controller; note that the voltage signal value is stable at 14 V due to battery and charge controller. The current starts at 1.5 A and the power value of 10 W at the start of operation of the experiment, and at the largest electric load, the current increases to 2.5 A and power 25 W. Figure 6.25 shows the signals on the terminals of the electric loads, voltage (V_{ac-rms}), current

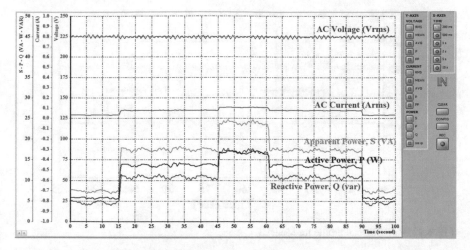

Fig. 6.20 Inverter output voltage, power, and current with battery

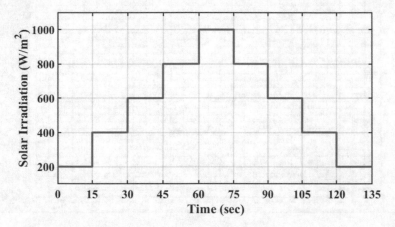

Fig. 6.21 Variable solar irradiation for Model 3

($I_{\text{ac} - \text{rms}}$), apparent power ($S$), active power ($P$), and reactive power ($Q$). Changes in AC current and powers are shown as a result of the enter and exit of electrical loads to adjust the output of the off-grid inverter according to the required load power and to maintain the constant voltage value at 225 V_{rms}.

Figure 6.26 presents the AC voltage and current waveform at enter and exit load. Figure 6.26a shows that when the load power increases, the current increases and the voltage decreases, and voltage then returns to settle again. The reverse occurs in Fig. 6.26b when an electric current is switched off, the current value decreases, and the voltage increases instantaneous value and then returns to settle again.

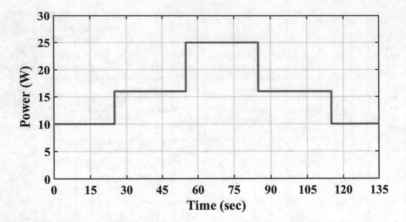

Fig. 6.22 Variable load profile for Model 3

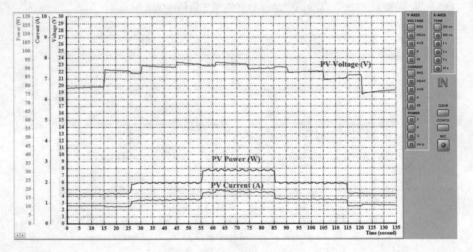

Fig. 6.23 PV output voltage, power, and current with battery and changing solar irradiation

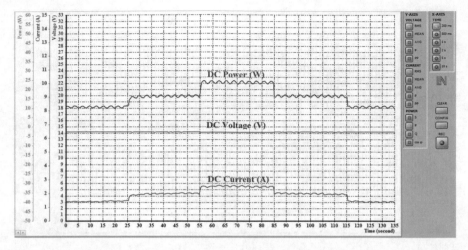

Fig. 6.24 Solar charge controller output voltage, power, and current with battery and changing solar irradiation

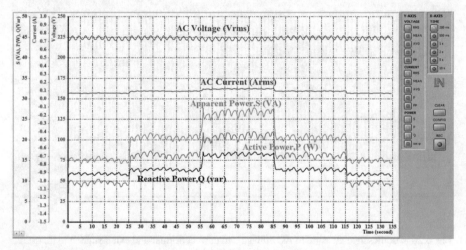

Fig. 6.25 Inverter output voltage, power, and current with battery and changing solar irradiation

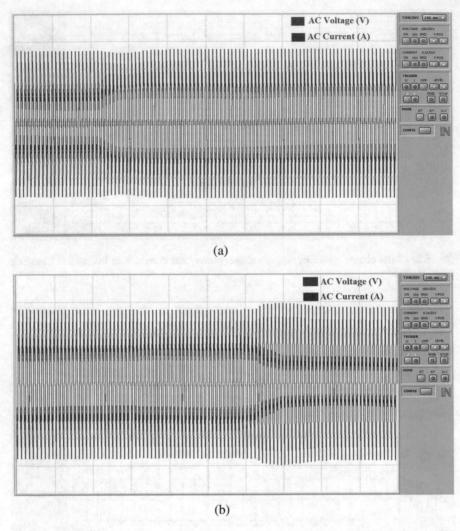

Fig. 6.26 Oscilloscope measure of AC voltage and current at enter and exit a sudden load. (**a**) AC voltage and current at enter a sudden load. (**b**) AC voltage and current at exit a sudden load

Chapter 7
Conclusions and Future Work

7.1 Conclusions

This book shed the light on the dynamic modeling, simulation, and control strategy of a stand-alone PV system with the energy storage system. The BS-HESS is considered a buffer store to eliminate the mismatch between power available from the PV array and power demand from the load. The BS-HESS is thus a practical solution to minimize the battery stress, battery size, and the total capital cost of the system. The main conclusions and recommendations drawn from this work are summarized next.

The first part of this work investigated the dynamic performance of the PV stand-alone system during variation of the environmental conditions. The effectiveness of the implemented MPPT techniques and the employed control strategy is evaluated during variations of the solar irradiance and the cell temperature.

1. The simulation results have verified the credibility of the implemented MPPT techniques in extraction the maximum power from the PV system during the rapid variation of the environmental conditions.
2. The introduced a review of two MPPT techniques that implemented in the PV systems, the perturb and observe MPPT Technique and Incremental Conductance MPPT technique.
3. The two MPPT techniques were simulated by the MATLAB/Simulink, and the results response of the PV array from voltage, current, and power are compared to the effect of solar radiation and temperature change.
4. The control strategy successfully keeps the load voltage constant regardless of the variations of the solar irradiance and temperature.

Then, the proposed PV stand-alone system is utilized to supply the demanded power of variable loads. This proposed PV stand-alone system consists of a PV array with a rated power capacity of 5 kW. The power flow control strategy is

© Springer Nature Switzerland AG 2019
A. A. Elbaset et al., *Performance Analysis of Photovoltaic Systems with Energy Storage Systems*, https://doi.org/10.1007/978-3-030-20896-7_7

proposed to feed the demanded power of the variable loads. From the simulation results of the studied cases, the following points may be concluded:

1. When the generated power from the PV system is greater than the demanded power of the variable loads, the surplus power will be injected into the BES through controlled bidirectional DC/DC converter act as a charge controller.
2. Otherwise, when the generated power from the PV stand-alone system is lower than the demanded power of the variable loads, the PV power system in cooperation with the BES will supply the variable loads.
3. Furthermore, the BES helps to improve the performance of the system through the control used in the process of charge/discharge to manage the sudden load changes and helps to maintain a stable voltage level on the load and PV terminals.

The third topic in this work was to improve the performance of the PV stand-alone system by leveraging the properties of the BS-HESS. This book proposed an efficient control strategy to enhance BS-HESS capable of the PV stand-alone system. A control strategy is essential for the BS-HESS to optimize the energy utilization and energy sustainability to a maximum extent as it is the algorithm which manages the power flow of the battery and SC. From the dynamic performance analysis of the PV stand-alone system with BS-HESS during the high fluctuation solar irradiation and variable load power for rural household load profile, the following points may be concluded:

1. Presented Simulink model of the proposed PV stand-alone system with BS-HESS and an optimal intelligent control strategy consist of the FLC and LPF and compare the intelligent strategy with the classic strategy which is based on the high pass filter.
2. The intelligent control strategy helps to reduce the dynamic stress and peak power demand of the battery by employing LPF and FLC, and the MFs of the FLC are implemented to optimize the battery peak current reduction.
3. The proposed system is evaluated and compared to the conventional system with battery-only systems and the systems with classic control strategies.
4. The simulation results show that the dynamic stress and peak current demand of the battery in the proposed system are greatly improved, which will eventually extend the battery life span.
5. The proposed system is able to operate the SC within the recommended SOC range and utilize the limited energy of SC effectively to perform better than the classic systems.

The main contributions of this work may be concluded in the following points:

• This book investigates the dynamic modeling, simulation, and control strategy of a PV stand-alone system.
• The performance of the PV stand-alone system is analyzed during variations of the solar irradiance and the cell temperature in order to evaluate the effectiveness of the implemented MPPT techniques by comparing the results of two techniques of MPPT (P&O and InCond).

- Also, in this work, the proposed PV stand-alone system with BES is utilized to supply the demanded power of variable loads at fluctuation solar irradiation.
- Moreover, the proposed PV stand-alone system with BS-HESS and its control strategy is proposed to feed the demanded power of the variable loads at high fluctuation solar irradiation.
- Furthermore, proposed intelligent control strategy to reduced battery dynamic stress and peak current demand of the battery, which will eventually extend the battery life span.
- Performance analysis of PV module with battery and variable AC loads which simulation at the experimental using MPPT have shown an accurate and the ability of the system to approach for MPP.
- Experimental results using a stand-alone PV system with different loads are more stable when using battery connected to system.

7.2 Suggestions for Future Work

Recently, the permanent growth of the energy demand and the rapid depletion of the conventional power sources have attracted the research interests toward the renewable energy sources especially the PV energy and wind energy as alternative sources of energy. The future researches in the renewable PV system and energy storage systems are suggested to be concentrated in the following research points:

- Integrate the PV system with wind power system and fuel cell to operate as PV/wind/fuel cell distributed generation system connected to the electrical grid.
- Optimal allocation of energy storage system for improving the performance of microgrid.
- Develop a manual guide for the type of batteries in PV stand-alone systems based on the nature of the electrical loads.

Appendices

Appendix A: Datasheet of 16 V Small Cell Module

FEATURES AND BENEFITS*
> 16V DC working voltage
> Resistive cell balancing
> Compact, light weight package
> Screw terminals

TYPICAL APPLICATIONS
> Wind turbine pitch control
> Small UPS systems

PRODUCT SPECIFICATIONS

ELECTRICAL	BMOD0058 E016 B02
Rated Capacitance[1]	58 F
Minimum Capacitance, initial[1]	58 F
Maximum Capacitance, initial[1]	70 F
Maximum ESR $_{DC}$ initial[1]	22 mΩ
Test Current for Capacitance and ESR$_{DC}$[1]	35 A
Rated Voltage	16 V
Absolute Maximum Voltage[2]	17 V
Absolute Maximum Current	170 A
Leakage Current at 25°C, maximum[3]	25 mA
Maximum Series Voltage	750 V
Capacitance of Individual Cells[9]	350 F
Maximum Stored Energy, Individual Cell[9]	0.35 Wh
Number of Cells	6

TEMPERATURE	
Operating Temperature (Cell Case Temperature)	
Minimum	-40°C
Maximum	65°C
Storage Temperature (Stored Uncharged)	
Minimum	-40°C
Maximum	70°C

© Springer Nature Switzerland AG 2019
A. A. Elbaset et al., *Performance Analysis of Photovoltaic Systems with Energy Storage Systems*, https://doi.org/10.1007/978-3-030-20896-7

PRODUCT SPECIFICATIONS (Cont'd)

PHYSICAL	BMOD0058 E016 B02
Mass, typical	0.63 kg
Power Terminals	M5 Thread
Recommended Torque - Terminal	4 Nm
Vibration Specification	IEC60068-2-6
Shock Specification	IEC60068-2-27, -29
Environmental Protection	IP54
Cooling	Natural Convection

MONITORING / CELL VOLTAGE MANAGEMENT	
Internal Temperature Sensor	N/A
Temperature Interface	N/A
Cell Voltage Monitoring	N/A
Connector	N/A
Cell Voltage Management	Passive

POWER & ENERGY	
Usable Specific Power, P_d[4]	2,200 W/kg
Impedance Match Specific Power, P_{max}[5]	4,600 W/kg
Specific Energy, E_{max}[6]	3.3 Wh/kg
Stored Energy, E_{stored}[7,9]	2.1 Wh

SAFETY	
Short Circuit Current, typical (Current possible with short circuit from rated voltage. Do not use as an operating current.)	730 A
Certifications	RoHS, UL810a (640 Volts)
High-Pot Capability[10]	5,600 VDC

TYPICAL CHARACTERISTICS

THERMAL CHARACTERISTICS	BMOD0058 E016 B02
Thermal Resistance (R_{ca}, All Cell Cases to Ambient), typical[8]	4.8°C/W
Thermal Capacitance (C_{th}), typical	420 J/°C
Maximum Continuous Current ($\Delta T = 15°C$)[8]	12 A$_{RMS}$
Maximum Continuous Current ($\Delta T = 40°C$)[8]	19 A$_{RMS}$

LIFE	
DC Life at High Temperature[1] (held continuously at Rated Voltage and Maximum Operating Temperature)	1,500 hours
Capacitance Change (% decrease from minimum initial value)	20%
ESR Change (% increase from maximum initial value)	100%
Projected DC Life at 25°C[1] (held continuously at Rated Voltage)	10 years
Capacitance Change (% decrease from minimum initial value)	20%
ESR Change (% increase from maximum initial value)	100%
Shelf Life (Stored uncharged at 25°C)	4 years

ESR AND CAPACITANCE VS TEMPERATURE

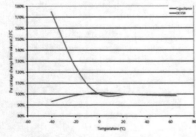

NOTES

1. Capacitance and ESR_{DC} measured at 25°C using specified test current per waveform below.
2. Absolute maximum voltage, non-repeated. Not to exceed 1 second.
3. After 72 hours at rated voltage. Initial leakage current can be higher.
4. Per IEC 62391-2, $P_d = \dfrac{0.12V^2}{ESR_{DC} \times mass}$
5. $P_{max} = \dfrac{V^2}{4 \times ESR_{DC} \times mass}$
6. $E_{max} = \dfrac{\frac{1}{2}CV^2}{3,600 \times mass}$
7. $E_{stored} = \dfrac{\frac{1}{2}CV^2}{3,600}$
8. $\Delta T = I_{RMS}^2 \times ESR \times R_{ca}$
9. Per United Nations material classification UN3499, all Maxwell ultracapacitors have less than 10 Wh capacity to meet the requirements of Special Provisions 361. Both individual ultracapacitors and modules composed of those ultracapacitors shipped by Maxwell can be transported without being treated as dangerous goods (hazardous materials) under transportation regulations.
10. Duration = 60 seconds. Not intended as an operating parameter.

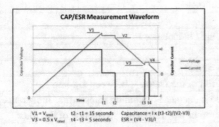

CAP/ESR Measurement Waveform

$V1 = V_{rated}$ t2 - t1 = 15 seconds   Capacitance = I x (t3-t2)/(V2-V3)
$V3 = 0.5 \times V_{rated}$ t4 - t3 = 5 seconds   ESR = (V4 - V3)/I

MOUNTING RECOMMENDATIONS

Recommended mounting screw M4. Maximum torque on mounting screws 4 Nm. All 6 mounting locations must be utilized to meet vibration specifications.

MARKINGS

Products are marked with the following information: Rated capacitance, rated voltage, product number, name of manufacturer, positive and negative terminal, and serial number.

BMOD0058 E016 B02

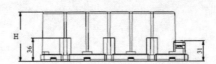

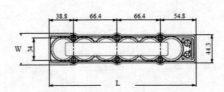

Part Description	Dimensions (mm)			Package Quantity
	L (±0.5mm)	W (±0.5m)	H (±0.5mm)	
BMOD0058 E016 B02	226.5	49.5	76.0	10

Appendix B: The M-File Program in MATLAB for Calculates the Values of LCL Filter Components

% System parameters	
$P_n = 5000$	% Inverter power: 5000 W
$E_n = 250$	% Grid voltage: 250 V
$V_{dc} = 280$	% DC link voltage: 280 V
$f_n = 60$	% Grid frequency: 60 Hz
$w_n = 2*p_i*f_n$	
$f_{sw} = 10,000$	% Switching frequency: 10000 Hz
$w_{sw} = 2*p_i*f_{sw}$	% Base values
$Z_b = (E_n{}^\wedge 2)/P_n$	
$C_b = 1/(w_n*Z_b)$	% Filter parameters
delta_Ilmax $= 0.1*((P_n*\mathrm{sqrt}(2))/E_n)$	
$L_i = V_{dc}/(16*f_{sw}*\mathrm{delta\_Ilmax})$	% Inverter side inductance
$x = 0.05$	
$C_f = x*C_b$	%Filter capacitor
$r = 0.6$	% Calculation of the factor, r, between L_{inv} and L_g
$L_g = r*L_i$	% Grid side inductance (including transformer inductance)
$w_{res} = \mathrm{sqrt}((L_i + L_g)/(L_i*L_g*C_f))$	% Calculation of w_{res}, resonance frequency of the filter
$f_{res} = w_{res}/(2*p_i)$	
$R_d = 1/(3*w_{res}*C_f)$	% Damping resistance

References

1. A.A. Elbaset, H. Ali, M. Abd-El Sattar, Novel seven-parameter model for photovoltaic modules. Sol. Energy Mater. Sol. Cells **130**, 442–455 (2014)
2. Y. Sawle, S.C. Gupta, A.K. Bohre, US solar photovoltaic manufacturing: Industry trends, global competition, federal support. Washington, DC Congr. Res. Serv. **6**, 11 (2012)
3. M.K. Hossain, M.H. Ali, Transient stability augmentation of PV/DFIG/SG-based hybrid power system by parallel-resonance bridge fault current limiter. Electr. Power Syst. Res. **130**, 89–102 (2016)
4. Y. Sawle, S.C. Gupta, A.K. Bohre, PV-wind hybrid system: A review with case study. Cogent Eng. **3**(1), 1–31 (2016)
5. A. Samadi, *Large Scale Solar Power Integration in Distribution Grids* (KTH Royal Institute of Technology, Stockholm, 2014)
6. D.C. Jones, Control Techniques for the Maximization of Power Converter Robustness and Efficiency in a Parallel Photovoltaic Architecture, Colorado, 2013
7. D. Feldman, G. Barbose, R. Wiser, N. Darghouth, A. Goodrich, Statistical Information about energy and renewable. Natl. Renew. Energy Lab. (November) (2012)
8. V. Rajasekaran, *Modeling, Simulation and Development of Supervision Control System for Hybrid Wind Diesel System* (Saint Mary's University, Halifax, 2013)
9. D. Sera, L. Mathe, T. Kerekes, S.V. Spataru, R. Teodorescu, On the perturb-and-observe and incremental conductance mppt methods for PV systems. IEEE J. Photovoltaics **3**(3), 1070–1078 (2013)
10. IEA-PVPS Reporting Countries, Becquerel Institute (BE), and RTS Corporation (JP), Snapshot of Global Photovoltaic Markets (1992–2016), IEA PVPS Task 1, International Energy Agency Power Systems Programme, Report IEA PVPS T1-31-2017, pp. 1–16, 2017
11. International Energy Agency, Technology roadmap for solar photovoltaic energy. Energy Technol. Perspect., 60 (2014)
12. S. of and G. Photovoltaic, The International Energy Agency (IEA) – Photovoltaic Power Systems Programme – 2018. Snapshot of Global Photovoltaic Markets, pp. 1–16, 2018
13. Egyptian government, Ministry of Electricity and Renewable Energy, New and Renewable Energy Authority. [Online]. Available: http://www.nrea.gov.eg/Technology/SolarIntro. Accessed Jan 03 2019
14. S. Sumathi, L. Ashok Kumar, P. Surekha, *Solar PV and Wind Energy Conversion Systems* (Springer, Cham, 2015)
15. M.M. Ahmed, M. Sulaiman, Design and proper sizing of solar energy schemes for electricity production in Malaysia, in *Power Engineering Conference, 2003. PECon 2003. Proceedings. National*, 2003, pp. 268–271

© Springer Nature Switzerland AG 2019
A. A. Elbaset et al., *Performance Analysis of Photovoltaic Systems with Energy Storage Systems*, https://doi.org/10.1007/978-3-030-20896-7

16. H.A. Khan, S. Pervaiz, Technological review on solar PV in Pakistan: Scope, practices and recommendations for optimized system design. Renew. Sust. Energ. Rev. **23**, 147–154 (2013)
17. E. Alsema, in *Practical Handbook of Photovoltaics*. Energy payback time and CO2 emissions of PV systems, (Academic Press, 2012), pp. 1097–111
18. S. Duryea, S. Islam, W. Lawrance, A battery management system for stand alone photovoltaic energy systems, in *Industry Applications Conference, 1999. Thirty-Fourth IAS Annual Meeting. Conference Record of the 1999 IEEE*, 1999, vol. 4, pp. 2649–2654
19. G. Boyle, *Renewable Energy*, ed. by G. Boyle, (Oxford University Press, 2004), p. 456
20. Y. Li, C. Chen, Q. Xie, Research of an improved grid-connected PV generation inverter control system, *2010 Int. Conf. Power Syst. Technol. Technol. Innov. Mak. Power Grid Smarter, POWERCON2010*, no. 1, pp. 2–7, 2010
21. IEEE Draft Guide for Array and Battery Sizing in Stand- Alone Photovoltaic Systems Prepared by the Energy Storage Subsystems Working Group of the IEEE Standards Coordinating Committee 21, Fuel Cells, Photovoltaics, Dispersed Generation, and Energy Storage Committee, in IEEE Std P1562/D8. (2006)
22. W. Song, W. Ren, H. Liu, Y. Zhao, L. Wu, An evaluation index system for hybrid wind/PV/ energy storage power generation system operating characteristics in multiple spatial and temporal scales. Int. Conf. Renew. Power Gener. (RPG 2015), 6 (2015)
23. H. Wang, Z. Jiancheng, Research on charging/discharging control strategy of battery-super capacitor hybrid energy storage system in photovoltaic system, in *Power Electronics and Motion Control Conference (IPEMC-ECCE Asia), 2016 IEEE 8th International*, 2016, pp. 2694–2698
24. O. Hazem Mohammed, Y. Amirat, M. Benbouzid, G. Feld, T. Tang, A.A. Elbaset, Optimal design of a stand-alone hybrid PV/fuel cell power system for the city of Brest in France. Int. J. Energy Convers. **2**(1), 1–7 (2014)
25. S. Pati, S.K. Kar, K.B. Mohanty, D. Panda, Voltage and frequency stabilization of a micro hydro –PV based hybrid micro grid using STATCOM equipped with battery energy storage system. IEEE Int. Conf. Power Electron. Drives Energy Syst. PEDES 2016. **2016**(January), 1–5 (2017)
26. A. McEvoy, T. Markvart, L. Castaner, Chapter IIB-2 – Batteries in PV systems, in *Practical Handbook of Photovoltaics: Fundamentals and Applications*, (Academic Press, Amsterdam/ Boston, 2012)
27. U. Sangpanich, A novel method of decentralized battery energy management for stand-alone PV-battery systems, in *Power and Energy Engineering Conference (APPEEC), 2014 IEEE PES Asia-Pacific*, 2014, pp. 1–5
28. B.P. Roberts, Sodium-Sulfur (NaS) batteries for utility energy storage applications, in *Power and Energy Society General Meeting-Conversion and Delivery of Electrical Energy in the 21st Century, 2008 IEEE*, 2008, pp. 1–2
29. G.J. May, A. Davidson, B. Monahov, Lead batteries for utility energy storage: A review. J. Energy Storage **15**, 145–157 (2018)
30. H. Ibrahim, A. Ilinca, J. Perron, Energy storage systems-characteristics and comparisons. Renew. Sust. Energ. Rev. **12**(5), 1221–1250 (2008)
31. The Photovoltaic Education Network, Operation of Lead Acid Batteries | PVEducation. [Online]. Available: https://www.pveducation.org/pvcdrom/lead-acid-batteries/operation-of-lead-acid-batteries. Accessed Jan 03 2019
32. T.B. Issa, P. Singh, M.V. Baker, T. Lee, Potentiometric measurement of state-of-charge of lead-acid batteries using polymeric ferrocene and quinones derivatives. J. Analytical Sci. Methods Instrum. **4**(December), 110–118 (2014)
33. D. Berndt, *Maintenance-Free Batteries: Lead-Acid, Nickel-Cadmium, Nickel-Hydride...* (Research Studies Press, Taunton, 1994)
34. J. Weniger, T. Tjaden, V. Quaschning, Sizing of residential PV battery systems. Energy Procedia **46**, 78–87 (2014)
35. NA, How to Design Solar PV System – Guide for sizing your solar photovoltaic system, *Leonics Co., Ltd.*, 2013. [Online]. Available: http://www.leonics.com/support/article2_12j/ articles2_12j_en.php. Accessed Jan 04 2019

36. F. Rafik, H. Gualous, R. Gallay, A. Crausaz, A. Berthon, Frequency, thermal and voltage super-capacitor characterization and modeling. J. Power Sources **165**(2), 928–934 (2007)
37. H. Gualous, D. Bouquain, A. Berthon, J.M. Kauffmann, Experimental study of supercapacitor serial resistance and capacitance variations with temperature. J. Power Sources **123**(1), 86–93 (2003)
38. A. Sumper, F. Díaz-González, O. Gomis-Bellmunt, *Energy Storage in Power Systems* (Wiley, New Delhi, 2016)
39. K. Ishaque, Z. Salam, H. Taheri, Simple, fast and accurate two-diode model for photovoltaic modules. Sol. Energy Mater. Sol. Cells **95**(2), 586–594 (2011)
40. J.K. Maherchandani, C. Agarwal, M. Sahi, Estimation of solar cell model parameter by hybrid genetic algorithm using Matlab. Int. J. Adv. Res. Comput. Eng. Technol. **1**(6), 78 (2012)
41. Z. Ahmad, S.N. Singh, Extraction of the internal parameters of solar photovoltaic module by developing Matlab/Simulink based model. Int. J. Appl. Eng. Res. **7**, 1 (2012)
42. S. Lineykin, M. Averbukh, A. Kuperman, Issues in modeling amorphous silicon photovoltaic modules by single-diode equivalent circuit. IEEE Trans. Ind. Electron. **61**(12), 6785–6793 (2014)
43. B.C. Babu, S. Gurjar, A novel simplified two-diode model of photovoltaic (PV) module. IEEE J. Photovoltaics **4**(4), 1156–1161 (2014)
44. N.D. Kaushika, N.K. Gautam, Energy yield simulations of interconnected solar PV arrays. IEEE Trans. Energy Convers. **18**(1), 127–134 (2003)
45. W. Xiao, W.G. Dunford, A modified adaptive hill climbing MPPT method for photovoltaic power systems, in *Power Electronics Specialists Conference, 2004. PESC 04. 2004 IEEE 35th Annual*, 2004, vol. 3, pp. 1957–1963
46. D. Sera, R. Teodorescu, J. Hantschel, M. Knoll, Optimized maximum power point tracker for fast changing environmental conditions. IEEE Int. Symp. Ind. Electron., 2401–2407 (2008)
47. K.-H. Chao, J.-P. Chen, A maximum power point tracking method based on particle swarm optimization for photovoltaic module arrays with shadows. ICIC Express Lett. **8**(1–5), 626–633 (2014)
48. G.-C. Hsieh, H.-I. Hsieh, C.-Y. Tsai, C.-H. Wang, Photovoltaic power-increment-aided incremental-conductance MPPT with two-phased tracking. IEEE Trans. Power Electron. **28**(6), 2895–2911 (2013)
49. S. Mekhilef, K.S. Tey, S. Mekhilef, S. Member, Modified incremental conductance algorithm for photovoltaic system under partial shading conditions and load variation modified incremental conductance algorithm for photovoltaic system under partial shading conditions and load variation. IEEE Trans. Ind. Electron. **61**(10), 5384–5392 (2014)
50. Q. Mei, M. Shan, L. Liu, J.M. Guerrero, A novel improved variable step-size incremental-resistance MPPT method for PV systems. IEEE Trans. Ind. Electron. **58**(6), 2427–2434 (2011)
51. T. Esram, J.W. Kimball, P.T. Krein, P.L. Chapman, P. Midya, Dynamic maximum power point tracking of photovoltaic arrays using ripple correlation control. IEEE Trans. Power Electron. **21**(5), 1282–1290 (2006)
52. A. Al Nabulsi, R. Dhaouadi, Efficiency optimization of a DSP-based standalone PV system using fuzzy logic and dual-MPPT control. IEEE Trans. Ind. Inf. **8**(3), 573–584 (2012)
53. E. Karatepe, T. Hiyama, Artificial neural network-polar coordinated fuzzy controller based maximum power point tracking control under partially shaded conditions. IET Renew. Power Gener. **3**(2), 239–253 (2009)
54. H. Renaudineau et al., A PSO-based global MPPT technique for distributed PV power generation. IEEE Trans. Ind. Electron. **62**(2), 1047–1058 (2015)
55. E. Bianconi et al., A fast current-based MPPT technique employing sliding mode control. IEEE Trans. Ind. Electron. **60**(3), 1168–1178 (2013)
56. F. Paz, M. Ordonez, Zero oscillation and irradiance slope tracking for photovoltaic MPPT. IEEE Trans. Ind. Electron. **61**(11), 6138–6147 (2014)
57. M. Killi, S. Samanta, Modified perturb and observe MPPT algorithm for drift avoidance in photovoltaic systems. IEEE Trans. Ind. Electron. **62**(9), 5549–5559 (2015)
58. R.M. Schupbach, J.C. Balda, Comparing DC-DC converters for power management in hybrid electric vehicles. IEMDC 2003 – IEEE Int. Electr. Mach. Drives Conf. **3**(C), 1369–1374 (2003)

59. F.A. Himmelstoss, M.E. Ecker, Analyses of a bidirectional DC-DC half-bridge converter with zero voltage switching, in *Signals, Circuits and Systems, 2005. ISSCS 2005. International Symposium on*, 2005, vol. 2, pp. 449–452
60. J. Cao, A. Emadi, A new battery/ultracapacitor hybrid energy storage system for electric, hybrid, and plug-in hybrid electric vehicles. IEEE Trans. Power Electron. **27**(1), 122–132 (2012)
61. W.L. Jing, C.H. Lai, W.S.H. Wong, D.M.L. Wong, Smart hybrid energy storage for stand-alone PV microgrid: Optimization of battery lifespan through dynamic power allocation. Appl. Mech. Mater. **833**, 19–26 (2016)
62. I. Shchur, Y. Biletskyi, Interconnection and damping assignment passivity-based control of semi-active and active battery/supercapacitor hybrid energy storage systems for stand-alone photovoltaic installations, in *Advanced Trends in Radioelecrtronics, Telecommunications and Computer Engineering (TCSET), 2018 14th International Conference on*, 2018, pp. 324–329
63. L.W. Chong, Y.W. Wong, R.K. Rajkumar, D. Isa, Modelling and simulation of standalone PV systems with battery-supercapacitor hybrid energy storage system for a rural household. Energy Procedia **107**(September 2016), 232–236 (2017)
64. A.A. Elbaset, H. Radwan, M.A. Sayed, G. Shabib, The non ideality effect of optimizing the P&O MPPT algorithm for PV AC load applications, in *17th International Middle East Power Systems Conference*, 2015
65. S. Nema, R.K. Nema, G. Agnihotri, Matlab/Simulink based study of photovoltaic cells/modules/array and their experimental verification. Int. J. Energy Environ. **1**(3), 487–500 (2010)
66. T. Salmi, M. Bouzguenda, A. Gastli, A. Masmoudi, Matlab/Simulink based modeling of photovoltaic cell. Int. J. Renew. Energy Res. **2**(2), 213–218 (2012)
67. Z.M. Karimi, Modelling, Implementation and Performance Analysis of a Hybrid Wind Solar Power Generator with Battery Storage, University of Coimbra, 2014
68. B.M. Hasaneen, A.A.E. Mohammed, Design and simulation of DC/DC boost converter, in *Power System Conference, 2008. MEPCON 2008. 12th International Middle-East*, 2008, pp. 335–340
69. D.W. Hart, *Power Electronics* (Tata McGraw-Hill Education, Valparaiso, 2011)
70. L.K. Letting, Modelling and Optimised Control of a Wind-Photovoltaic Microgrid with Storage, Tshwane University of Technology, 2013
71. S.S. Dessouky, A.A. Elbaset, A.H.K. Alaboudy, H.A. Ibrahim, S.A.M. Abdelwahab, L.K. Letting, Performance improvement of a PV-powered induction-motor-driven water pumping system. 2016 18th Int. Middle-East Power Syst. Conf. MEPCON 2016 – Proc. (1), 373–379 (2013)
72. S. Bae, A. Kwasinski, Dynamic modeling and operation strategy for a microgrid with wind and photovoltaic resources. IEEE Trans. Smart Grid **3**(4), 1867–1876 (2012)
73. T. Esram, P.L. Chapman, Comparison of photovoltaic array maximum power point tracking techniques. IEEE Trans. Energy Convers. **22**(2), 439–449 (2007)
74. J. Jana, H. Saha, K. Das Bhattacharya, A review of inverter topologies for single-phase grid-connected photovoltaic systems. Renew. Sust. Energ. Rev. **72**, 1256–1270 (2017)
75. H.-G. Jeong, K.-B. Lee, S. Choi, W. Choi, Performance improvement of LCL-filter-based grid-connected inverters using PQR power transformation. IEEE Trans. Power Electron. **25**(5), 1320–1330 (2010)
76. M. Hanif, V. Khadkikar, W. Xiao, J.L. Kirtley, Two degrees of freedom active damping technique for LCL filter-based grid connected PV systems. IEEE Trans. Ind. Electron. **61**(6), 2795–2803 (2014)
77. C. Bao, X. Ruan, X. Wang, W. Li, D. Pan, K. Weng, Step-by-step controller design for LCL-type grid-connected inverter with capacitor–current-feedback active-damping. IEEE Trans. Power Electron. **29**(3), 1239–1253 (2014)
78. J. Lettl, J. Bauer, L. Linhart, Comparison of different filter types for grid connected inverter, in *PIERS Proceedings, Marrakesh, Morocco*, pp. 1426–1429, 2011

79. A. Reznik, M.G. Simões, A. Al-Durra, S.M. Muyeen, LCL filter design and performance analysis for grid-interconnected systems. IEEE Trans. Ind. Appl. **50**(2), 1225–1232 (2014)
80. B.S. Borowy, Z.M. Salameh, Methodology for optimally sizing the combination of a battery bank and PV array in a wind/PV hybrid system. IEEE Trans. Energy Convers. **11**(2), 367–375 (1996)
81. MathWorks, Implement generic battery model – Simulink. [Online]. Available: https://www.mathworks.com/help/physmod/sps/powersys/ref/battery.html?searchHighlight=Theequivalent circuitofbattery&s_tid=doc_srchtitle. Accessed Jan 05 2019
82. F. Caricchi, F. Crescimbini, F.G. Capponi, L. Solero, Study of bi-directional buck-boost converter topologies for application in electrical vehicle motor drives, in *Applied Power Electronics Conference and Exposition, 1998. APEC'98. Conference Proceedings 1998., Thirteenth Annual*, 1998, vol. 1, pp. 287–293
83. H.R. Karshenas, H. Daneshpajooh, A. Safaee, P. Jain, A. Bakhshai, Bidirectional dc-dc converters for energy storage systems, in *Energy Storage in the Emerging Era of Smart Grids*, (InTech, Rijeka, 2011)
84. D. Committee, I. Power, E. Society, IEEE recommended practice and requirements for harmonic control in electric power systems. IEEE Power Energy Soc. Spons. **IEEE Std 5** (2014)
85. H. Tao, S. Liu, Design and implementation of digital control of photovoltaic power inverter. Procedia Environ. Sci. **11**(Part A), 155–162 (2011)
86. A.Q. Jakhrani, A.R.H. Rigit, A.-K. Othman, S.R. Samo, S.A. Kamboh, Life cycle cost analysis of a standalone PV system, in *Green and Ubiquitous Technology (GUT), 2012 International Conference on*, 2012, pp. 82–85
87. S.Y. Kan, M. Verwaal, H. Broekhuizen, The use of battery–capacitor combinations in photovoltaic powered products. J. Power Sources **162**(2), 971–974 (2006)
88. A.C. Baisden, A. Emadi, ADVISOR-based model of a battery and an ultra-capacitor energy source for hybrid electric vehicles. IEEE Trans. Veh. Technol. **53**(1), 199–205 (2004)
89. H. Zhou, T. Bhattacharya, D. Tran, T.S.T. Siew, A.M. Khambadkone, Composite energy storage system involving battery and ultracapacitor with dynamic energy management in microgrid applications. IEEE Trans. Power Electron. **26**(3), 923–930 (2011)
90. M.E. Glavin, P.K.W. Chan, S. Armstrong, W.G. Hurley, A stand-alone photovoltaic supercapacitor battery hybrid energy storage system, in *Power Electronics and Motion Control Conference, 2008. EPE-PEMC 2008. 13th*, 2008, pp. 1688–1695
91. R.A. Dougal, S. Liu, R.E. White, Power and life extension of battery-ultracapacitor hybrids. IEEE Trans. Compon. Packag. Technol. **25**(1), 120–131 (2002)
92. MathWorks, Implement generic supercapacitor model – Simulink. [Online]. Available: https://www.mathworks.com/help/physmod/sps/powersys/ref/supercapacitor.html. Accessed Jan 05 2019
93. F. Garcia-Torres, C. Bordons, Optimal economical schedule of hydrogen-based microgrids with hybrid storage using model predictive control. IEEE Trans. Ind. Electron. **62**(8), 5195–5207 (2015)
94. N. Mendis, K.M. Muttaqi, S. Perera, Active power management of a super capacitor-battery hybrid energy storage system for standalone operation of DFIG based wind turbines. Conf. Rec. – IAS Annu. Meet. IEEE Ind. Appl. Soc., 1–8 (2012)
95. L.W. Chong, Y.W. Wong, R.K. Rajkumar, D. Isa, An optimal control strategy for standalone PV system with Battery- Supercapacitor Hybrid Energy Storage System. J. Power Sources **394**, 35–49 (2018)
96. E. Zhandire, Solar resource classification in South Africa using a new index. J. Energy South. Afr. **28**(?), 61–70 (2017)
97. M.I. Fahmi, R.K. Rajkumar, R. Arelhi, D. Isa, Study on the effect of supercapacitors in solar PV system for rural application in Malaysia, in *Power Engineering Conference (UPEC), 2015 50th International Universities*, 2015, pp. 1–5

Index

Printed in the United States
By Bookmasters